# Recent Developments in Nanofibers Research

*Edited by Maaz Khan*
*and Samson Jerold Samuel Chelladurai*

Published in London, United Kingdom

Recent Developments in Nanofibers Research
http://dx.doi.org/10.5772/intechopen.100676
Edited by Maaz Khan and Samson Jerold Samuel Chelladurai

Contributors
Rebika Baruah, Archana Moni Das, Dheeraj Kumar, Nadeem Faisal, Swarup S. Deshmukh, Ramit Choudhury, Gurpreet Singh Matharou, Simran Kaur, Alena Opálková Šišková, Heba M. Abdallah, Smaher Mosad Elbayomi, Anita Eckstein Andicsová, Govind Kumar Sharma, Nirmala Rachel James, Vijayakumar Chinnaswamy Thangavel, Shamim Rishwana Syed Mohammed, Arunjunai Raj Mahendran, Siva Kaylasa Sundari Saravanamuthu



First published in London, United Kingdom, 2023 by IntechOpen
IntechOpen is the global imprint of INTECHOPEN LIMITED, registered in England and Wales, registration number: 11086078, 5 Princes Gate Court, London, SW7 2QJ, United Kingdom

British Library Cataloguing-in-Publication Data
A catalogue record for this book is available from the British Library

Additional hard and PDF copies can be obtained from orders@intechopen.com

Recent Developments in Nanofibers Research
Edited by Maaz Khan and Samson Jerold Samuel Chelladurai
p. cm.
Print ISBN 978-1-80356-386-2
Online ISBN 978-1-80356-387-9
eBook (PDF) ISBN 978-1-80356-388-6

# Meet the editors

Dr. Maaz Khan is a Deputy Chief Scientist at PINSTECH, Pakistan. He completed his Ph.D. and post-doctoral research in the field of material science (nanoscience). His research interests include the fabrication of nanomaterials and their structural, optical, magnetic, and electrical characterizations. He has authored more than 100 research articles and published eleven books. He is currently Editor-in-Chief of *The Nucleus and Journal of Materials, Processing and Design*, and Executive Editor of the *International Journal of Nano Studies and Technology.* He serves as an editorial board member of several other materials science journals.

Dr. Samson Jerold Samuel Chelladurai is an Associate Professor in the Department of Mechanical Engineering, Sri Krishna College of Engineering and Technology (SKCET), India. He obtained a BE degree in Mechanical Engineering and an ME in Manufacturing Engineering from Anna University, Chennai, India. He received a Ph.D. in Composite Materials from Anna University in 2018. He specializes in the manufacturing of metal matrix composites, characterization of composites, tribology, and optimization techniques. He has published many journal articles and five books. He has published five patents and received grants worth USD 89,000. He has received several awards from the Sri Krishna College of Engineering and Technology (SKCET) and other organizations.

# Contents

# Preface

Nanofibers are one-dimensional fibers with diameters in the nanometer range that can be generated from different polymers. They have different physical and chemical properties, and are fabricated using various techniques such as electrospinning, self-assembly, template-assisted synthesis, and thermal-induced phase separation. This book presents an overview of the current status of nanofibers, their fabrication and their functional applications in various fields.

The six chapters of the book, written by expert authors, provide the reader with a clear understanding of the electrospinning process, its theory, methodology, materials, and applications. Appropriate references at the end of each chapter lead readers to the best sources in the literature for a more detailed study.

I am indebted to all the authors for helping me to complete this project and to the whole IntechOpen publishing team for making the project possible. I am especially grateful to Ms. Karla Skuliber, Author Service Manager, for her cooperation during the publishing process.

**Dr. Maaz Khan**
Deputy Chief Scientist,
Pakistan Institute of Nuclear Science and Technology,
Islamabad, Pakistan

**Samson Jerold Samuel Chelladurai**
Associate Professor,
Department of Mechanical Engineering,
Sri Krishna College of Engineering and Technology,
Coimbatore, India

Chapter 1

# Electrospinning: The Technique and Applications

*Govind Kumar Sharma and Nirmala Rachel James*

## Abstract

Electrospinning is a useful and convenient method for producing ultrathin fibers. It has grabbed the scientific community's interest due to its potential to produce fibers with various morphologies. Numerous efforts have been made by researchers and industrialists to improve the electrospinning setup and the associated techniques in order to regulate the morphology of the electrospun fibers for practical applications. Porous, hollow, helical, aligned, multilayer, core-shell, and multichannel fibers have been fabricated for different applications. This chapter aims to provide readers with a clear understanding of the electrospinning process: its principle, methodology, materials, and applications. The chapter begins with a brief introduction to the history of electrospinning, followed by a discussion of its principle and the basic components of electrospinning setup. The parameters that affect the electrospinning process such as operating parameters and the properties of the material being electrospun are discussed briefly. An overview of the different types of electrospinning technique, capable of producing nanofibers with different morphologies, is also presented. Afterward, the applications of electrospun nanofibers, including their use in biomedical applications, filtration, energy sectors, and sensors applications are discussed succinctly. The perspectives on the challenges, opportunities, and new directions for future development of electrospinning technology are also offered.

**Keywords:** electrospinning, electrospun fiber, coaxial electrospinning, tri-axial electrospinning, core-shell fiber, spinneret

## 1. Introduction

Electrospinning (electrostatic fiber spinning) has been developed as a sophisticated, modern, and versatile technique since the late 1990s due to its ease of generating nanofibers with a range of materials. It uses an electric field to produce microscopic threads with diameters as small as nanometers (nm). Tissue engineering, filtration, energy, biotechnology, and sensors are just a few of the fields where electrospun fiber membranes find extensive applications [1–5].

### 1.1 History of electrospinning

In nature, fibers can be found in the shape of elongated objects or continuous filaments. Spiders have relied on webs of fiber matting to catch food in the wild. Silk fibers with sizes ranging from 2 to 5 meters make up the webs. Silkworms are also known for their capacity to produce silk strands in large quantities. Rayon is the name given to the earliest man-made textiles created from cotton or wood cellulose fibers. DuPont developed nylon as the first commercially feasible synthetic fabric in 1938, and it immediately sparked a popular interest [6–10].

A variety of techniques have been used to manufacture synthetic polymer fibers. The most common procedures are wet, dry, melt, and gel spinning. During the wet spinning process, a spinneret is immersed in a chemical bath. A polymer solution is extruded from a spinneret into a chemical bath, and then the polymer is precipitated out due to the chemical reaction or dilution effect to generate fibers through solidification. During dry spinning, a polymer solution is extruded into the air through a spinneret, and fibers are generated as a result of solvent evaporation from jets aided through a stream of hot air. For melt spinning, a polymer melt is extruded from the spinneret to produce fibers upon cooling. Gel spinning is used to generate fibers with high mechanical strength or other distinctive properties through spinning a polymer in the gel state followed by drying in air and then cooling in a liquid bath. In the gel spinning process, jets are mainly made under external shearing forces and/or mechanical drawing while passing through spinnerets, and fibers are generated upon solidification of the jets as a result of drying or precipitation. The fibers obtained by spinning have diameters in the range of 10–100 micrometers (μm) [11–15].

In 1887, fibers were made from a viscoelastic liquid in the presence of an external electric field as reported by Charles V boys. He used a setup that consisted of an insulated dish connected with an electric supply. He demonstrated that viscoelastic liquid could be drawn into fibers when it moved to the edge of the dish [16]. Electrospinning is a well-known technique now, for the production of continuous ultrathin fibers with diameters ranging from 10 nm to 100 μm. In 1600, William Gilbert introduced the concept of electrospinning. In his study, cone-shaped water droplet formation was observed in the presence of an electric field [4]. In 1747, Abbe Nollet demonstrated the earliest known electrospraying experiment, in which water could be sprayed as aerosol while passing through an electrostatically charged container that was placed on the ground. Both electrospinning and electrospraying are dependent on the use of a high voltage to eject viscous and viscoelastic liquid jets. During electrospinning, the jets eject polymer solution continuously to produce fibers instead of breaking into droplets as with electrospraying. In 1902, John Cooley and William Morton were granted two patents on electrospinning, in which they described a prototype of an electrospinning setup [4, 17]. In 1914, John Zeleny had reported that he was working on the treatment of liquid drop at the end of capillaries. In this study, he tried to find a mathematical model of liquids under electrostatic forces. Formhals tried to produce electrospun fibers in the 1930s, but the system had some disadvantages such as drying, due to the distance between nozzle and collector [18, 19]. In 1940, he improved and modified the device to overcome the drawback. Between 1964 and 1969, Sir Geoffrey Ingram Taylor developed the theoretical underpinning of electrospinning, and his research helped advancement in electrospinning by modeling the hopper form in which liquid drops were formed by the electric field. His collaboration with JR Melcher led to expanding the "leaky dielectric model" for conducting liquids. In early 1990s, some research groups, remarkably those led by Darrel Reneker and Gregory Rutledge, reinvented the electrospinning technique. These groups demonstrated that many different organic polymers could be electrospun into the nanoscale. For the first time, Darrel Reneker used high voltage to charge the polymer dispersion to produce fine fibers with a diameter of less than 5 μm [4, 20–22].

This technique became a very popular and a good choice for producing continuous and long fibers with diameters ranging from micrometer to nanometer. Meanwhile, new methods were developed to control the alignment and structure of electrospun nanofibers, which created new opportunities in energy-related and biomedical applications. After that, many other new methods were developed for aligning the nanofibers to improve several properties of nanofibers such as size, structure, morphology, composition, porosity, and conductivity. One such method is coaxial electrospinning to produce continuous core-sheath and hollow nanofibers [2, 23–27].

## 2. Electrospinning setup

Electrospinning is a voltage-driven technology that uses an electro-hydrodynamic process, in which a high voltage is applied to the polymer solution and then a liquid droplet is electrified to generate a jet, followed by elongation and stretching to produce fibers. The diameter of these fibers ranges from nanometers to a few micrometers (μm). One of the primary advantages of electrospinning is its adaptability in processing, which allows it to create fibers with a variety of configurations and morphological structures [19, 28]. **Figure 1** represents the schematic diagram of the basic electrospinning set-up. The basic electrospinning setup consists of mainly three parts: (1) a high voltage power supply, (2) a spinneret (metallic needle), and (3) a grounded collector [7, 29–31]. The metallic collectors are generally of three types, namely stationary flat plates, spinning drums, and rotating disc, as shown in **Figure 2**.

### 2.1 Principle

To understand the basic principle of the electrospinning process, consider a spherically charged droplet of a low-molecular-weight conducting liquid that is placed in a vacuum. The liquid droplet experiences two forces: (1) disintegrative repulsive

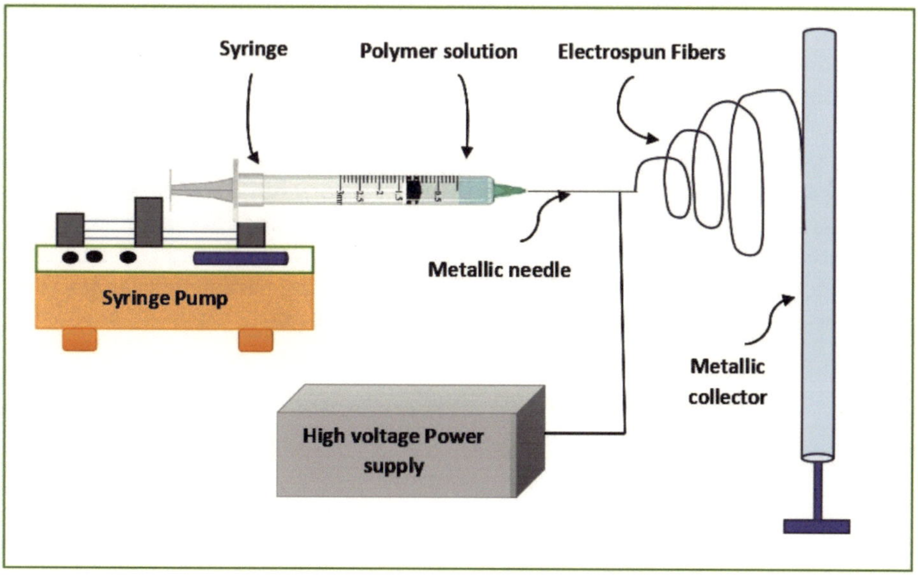


**Figure 1.**
*Basic setup of electrospinning [28].*

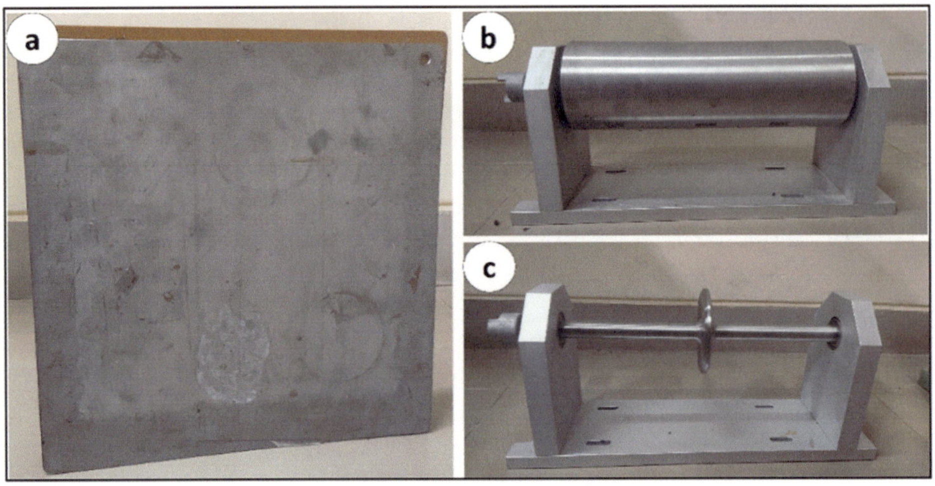


**Figure 2.**
*Metallic collectors. (a) Stationary flat plate, (b) drum collector, and (c) rotating disc.*

force and (2) surface tension that tries to hold the liquid droplet in a spherical shape. During the electrospinning process, a high electric voltage is applied to the liquid droplet from polymer melt/solution at the tip of the spinneret. When the high voltage is continuously increased, the liquid droplet will start to elongate into a conical shape known as a "Taylor cone." The elongation starts when the electrostatic repulsion overcomes surface tension. The charged liquid jet is directed toward the metallic collector once the Taylor cone has been produced. The liquid mentioned here can be melt polymer, polymer solution, or an emulsion. Solid fibers will develop as the melt cools down or solvent evaporates from the whipping action that happens throughout the flight time from the Taylor cone to the collector, depending on the liquid viscosity. As a result, the collector is covered with a non-woven fiber mat [32–35].

## 3. Parameters that affect the electrospinning process

The electrospinning process depends on operating parameters, material parameters, and ambient parameters that affect the morphology of fiber. The operating parameters consist of the applied voltage or electric field, the flow rate of polymer melt/solution, the distance between the tip of the metallic needle and collector, and the diameter of the needle. A small change in the operating parameters can lead to a significant change in the morphology of fiber [23, 26, 36–39].

### 3.1 Operating parameters

#### *3.1.1 Applied voltage*

The applied voltage determines the amount of charges carried by the jet, the degree of electrostatic repulsion among the charges, and the strength of interactions between the jet and the external electric field. Higher voltage facilitates the formation of thinner fibers, but it can also result in more fluid being ejected, resulting in thicker diameter fibers [36, 40].

#### *3.1.2 Flow rate*

It is necessary to adjust the flow rate of the spinning liquid for a particular voltage in order to maintain a stable Taylor cone during the electrospinning process. A uniform Taylor cone can produce uniform fibers with narrow dispersion during electrospinning. As the flow rate increased, the amount of material passing through the tip increased, resulting in the formation of fiber with a high diameter. At a very high flow rate, the polymeric jet becomes unstable due to the effect of gravitational force and tends to electrospray [40].

#### *3.1.3 The distance between the tip of the metallic needle and the collector*

The distance between the tip and the collector can also influence the diameters and shape of nanofibers; however, the effect is not as strong as the other factors. In the electrospinning process, a minimum distance is necessary to allow enough time for solvent evaporation before the fiber reaches the collector. Thinner fibers have resulted from longer distances. When the distance was too great or too small, beads would form [4, 7, 25].

#### *3.1.4 Diameter of the needle*

Nanofibers, electrospun with small needle diameters, are thinner, smoother, and bead-free and have greater fiber porosity than nanofibers electrospun with large needle diameters. As aforementioned, higher applied voltage, smaller spinneret diameter, and lower flow rate resulted in thinner electrospun nanofibers [4, 33, 36].

### 3.2 Material parameters

The material properties that affect the electrospinning process and fiber morphology involve polymer concentration, the viscosity of the solution, the surface tension of the polymer melt/solution, and other properties related to the solvent as well as the polymer itself [4, 7, 25].

#### *3.2.1 Polymer concentration*

Among the material properties, the polymer concentration plays the most significant role in stabilizing the fibrous structure because it influences the other properties such as viscosity of the solution, surface tension, and conductivity of the material [4, 7, 25].

#### *3.2.2 Viscosity*

The viscosity of the polymer solution has a significant impact on the diameter and shape of the electrospun fiber. It is controlled by polymer properties including molecular weight and polymer solution concentration. When the concentration of polymer in a solution is raised, the viscosity of the solution rises. If the viscosity is too high, then it will be difficult to pump the solution via the syringe pump, or the solution may dry at the needle tip before electrospinning can begin. Higher viscosity results in the increased diameter fibers and lower deposition region [4, 7, 25].

*3.2.3 Surface tension*

Surface tension is the attraction between molecules in a liquid that is influenced by intermolecular interactions. During electrospinning, the charges on the polymer solution must be high enough to overcome the solution's surface tension. When a high voltage is applied, the polymer jet begins to form from the needle's tips and elongates and stretches toward the collector, breaking up into minute droplets due to the solution's lower surface tension, which is known as electrospraying. Surface tension causes bead formation when the polymer concentration is low. If the surface tension is low, the formation of the jet begins at a lower voltage. The surface tension of polymer solution can be changed by varying solvents and by adding surfactants [26, 41].

*3.2.4 Conductivity*

During electrospinning, the charged liquid (melt/solution) stretches to form fibers due to charge repulsion. The jet carries more charges as the electrical conductivity of the solution rises, reducing the diameter of the electrospun fiber. A small amount of polyelectrolyte (salts) can be introduced to eliminate fiber bead formation since it raises more charges and helps to elongate the jet to produce fibers. The electrospinning of polymer solution is difficult at very high voltages, while fiber formation is impossible when the solution has no conductivity [4, 7, 25].

*3.2.5 Solvent properties*

For electrospinning, solvent choice is critical, since the process is influenced by the solvent's evaporation rate (which is determined by the solvent's vapor pressure) and permittivity [4, 7, 25].

*3.2.5.1 Vapor pressure*

The volatility or vapor pressure of the solvent determines the evaporation rate and, as a result, the solidification rate of the jet. High volatility is not suitable for spinning fibers because the jet may solidify immediately after leaving the spinneret. If the volatility is too low, the fibers will still be wet when deposited onto the collector. The solvent volatility modifies the surface fiber morphology and nano-membrane structure [4, 7, 25].

*3.2.5.2 Permittivity*

The permittivity of a solvent has a substantial impact on the electrospinning process and fiber morphology. The bead formation and diameter of the electrospun fiber are reduced when a solution with a greater permittivity is used. Higher permittivity increases bending instability and the traversed jet path of the electrospinning jet, resulting in smaller fiber diameter and a larger deposition area. Solvents such as N, N-dimethylformamide (DMF) can be used to increase the permittivity of polymer solutions [4, 7, 25].

### 3.3 Ambient parameters

The interaction between the surrounding environment such as the temperature and humidity of the surrounding and the electrospinning jet may alter the electrospinning process and fiber morphology. The temperature is inversely proportional to the viscosity of the solution. So, the temperature may affect the properties of the electrospinning solution. Humidity may affect the porosity of fiber because humidity influences solvent evaporation [4, 7, 25].

## 4. Types of electrospinning

Electrospinning technique is divided into two categories based on the electrospinning setup: needle-based electrospinning and needleless electrospinning. Furthermore, needle-based electrospinning is classified into electrospinning using single nozzle, coaxial, tri-axial, and multichannel spinnerets. Electrospinning may also be categorized based on the state of the spinning material. Thus it is divided into three categories: melt electrospinning, emulsion electrospinning, and solution electrospinning.

### 4.1 Needle-based electrospinning

In needle electrospinning, a needle-like spinneret is utilized and a sharp "cone shape" arises at the needle tip under the influence of the electric field.

#### *4.1.1 Single nozzle*

Single nozzle electrospinning is the simplest and basic form of the technique in which only one needle is used as the spinneret as shown in **Figure 1**.

**Side-by-side electrospinning** is a modification of the basic single nozzle electrospinning. The side-by-side electrospinning process employs two parallel syringes to produce fibers with Janus structure as shown in **Figure 3**. The same voltage is applied to both solutions, and the fiber is usually separated due to repulsion between the two solutions. Producing a Janus structure is extremely difficult due to repulsion, and as a result, just a few studies have been conducted on this topic.

#### *4.1.2 Coaxial electrospinning*

A coaxial needle, made up of two concentric hollow needles, is used to produce a coaxially electrified jet for coaxial electrospinning. A simple approach to fabricate a coaxial needle is to insert a small (inner) needle into a larger (outer) needle in a coaxial

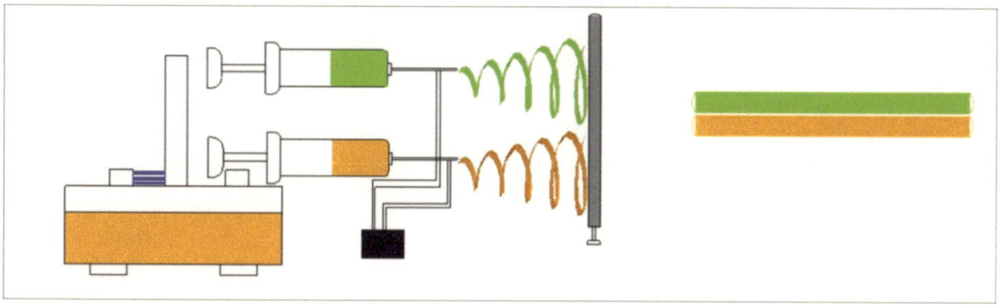

**Figure 3.**
*Side-by-side electrospinning setup.*

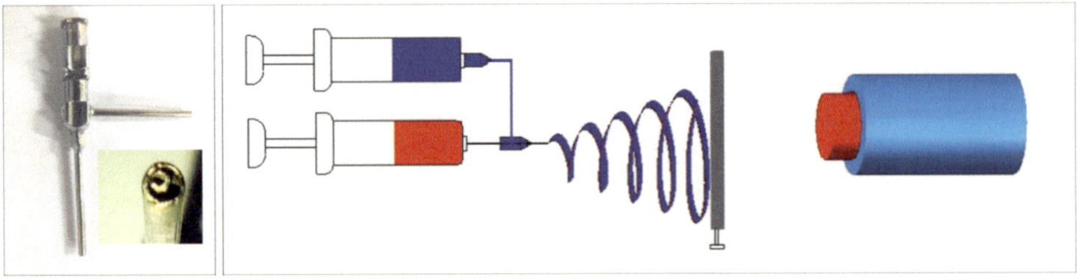

**Figure 4.**
*Coaxial spinneret and coaxial electrospinning setup.*

configuration. Using two syringe pumps, the outer and inner needles are subsequently filled with two solutions with independently controllable flow rates. When the core and shell solutions meet at the coaxial needle's exit end, in the presence of an external electric field, the shell solution wraps around the core solution to produce a compound Taylor cone, followed by the ejection of a coaxial jet. Finally, core-shell nanofibers with different core and shell compositions were fabricated, as shown in **Figure 4**.

In the fabrication of core-shell nanofibers, the properties of the inner and outer solutions, as well as the electrospinning parameters, all play important roles. For maintaining the smooth flow of jets, the inner and outer solutions, in particular, should have proper viscosities. Furthermore, the flow rates of the two solutions must be carefully controlled to ensure that the inner solution is completely wrapped by the outside solution. The flow rates may also be adjusted to change the diameter of the nanofibers and the shell thickness.

Coaxial electrospinning is used to fabricate core-shell nanofibers with better control over the compositions for different applications. It also allows for the fabrication of nanofibers from unspinnable liquids, as they may be used as the inner fluid although being controlled by the outer fluid. Hollow nanofibers with adjustable wall thickness can be fabricated by selectively removing the core from as-spun core-shell nanofibers. The core-shell morphology of coaxial electrospun fibers offers the possibility of multifunctional materials, with various functional components put into the two compartments of the concentric structure [24, 34, 42–54].

### *4.1.3 Tri-axial electrospinning*

The trilayer spinneret and tri-axial electrospinning are shown in **Figure 5**. It consists of three concentric metal capillaries. Tri-axial electrospinning is often used to construct a three-layer system with a drug-loaded core, a hydrophobic middle layer, and a hygroscopic exterior layer, as seen in the **Figure 5**. It is vital to avoid merging between the spinning solutions in order to generate high-quality multi-compartment fibers. This means that either the compartmentalized solutions must be immiscible or all of the solutions must evaporate the solvent at the same rate: if one of the liquids evaporates quicker than the others, the compartments will separate. So, the spinneret needs to be designed according to the application. A properly designed spinneret may be utilized to control the behavior of fluids in an electrical field as well as act as a template for constructing the right nanofiber structures [11, 55–57]. Lallave and coworkers [58] were the first to report tri-axial electrospinning. They employed a tri-axial arrangement of ethanol, lignin, and glycerine (from outermost to innermost layer) with ethanol sheath flow to avoid solidification of the Taylor cone and glycerine as a template fluid.

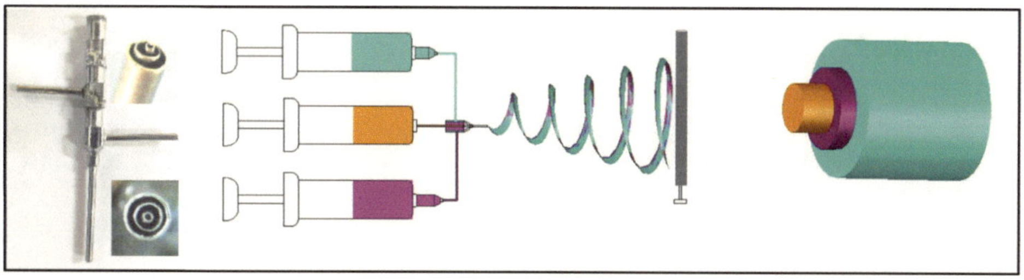

**Figure 5.**
*Tri-axial spinneret and tri-axial electrospinning setup.*

### *4.1.4 Multichannel electrospinning*

In the simplest example of multichannel electrospinning, three metallic capillaries are inserted in a syringe at the three vertices of an equilateral triangle as shown in **Figure 6**. Zhao et al. [59] for the first time presented a multifluid compound jet electrospinning approach that can easily and quickly produce bio-mimic hierarchical multichannel microtubes. Multi-axial electrospinning was used to create a biomimetic system. They used numerous inner axial paraffin oils in a $Ti(OiPr)_4$ solution, then removed the organics to construct a multilayer channel.

## 4.2 Needleless electrospinning

In needleless electrospinning, several cones are spun without the need for a needle or a tiny open structure. The jet initiation in needleless electrospinning is a self-organized process that happens on a free liquid surface and is not driven by capillary forces. It is based on the use of an external agitation force to concentrate the electric field on the free liquid surface to the intensity necessary to initiate a Taylor cone [60–63].

## 4.3 Melt electrospinning

Solvent removal, recycling, environmental concerns, and toxicity associated with the usage of solvents are all avoided with melt electrospinning. The melted polymer is injected into the capillary tube. The operation must be carried out in a vacuum, therefore the capillary tube, the charged melt fluid jet's passage, and the metal collector must all be enclosed in a vacuum. Melt electrospinning has the advantage of producing extremely homogeneous fibers with very little variation in fiber diameter. Aside from these benefits, there are drawbacks due to the

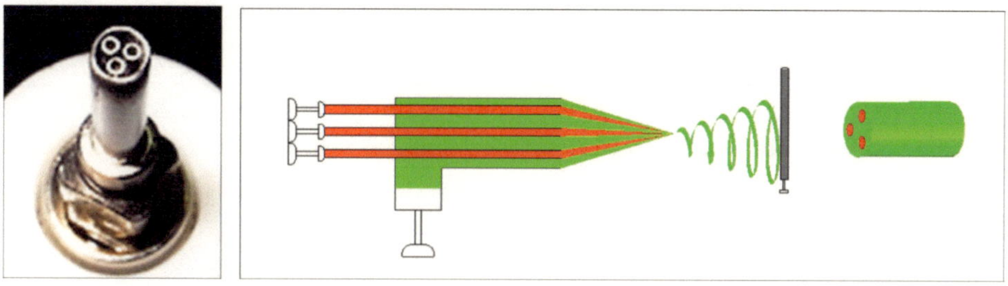

**Figure 6.**
*Multichannel spinneret and multichannel electrospinning setup.*

specialized equipment required, as well as the high viscosity and low electrical conductivity of polymer melt. In comparison to electrospinning from a polymer solution, despite the benefits afforded by melt electrospinning, the technology has not achieved more popularity and has not been frequently employed. This is mostly owing to the high viscosity, high process temperatures, and the inability to create nanometer-sized fibers [64–70].

### 4.4 Emulsion electrospinning

Emulsion electrospinning is a word that refers to two types of electrospinning. (1) the electrospinning of an emulsion through a single spinneret, whereby the emulsion structure allows reorganization to form a core-shell fiber, similar to coaxial electrospinning; (2) the electrospinning of an emulsion through multiple spinnerets, whereby the emulsion structure allows reorganization to form a core-shell fiber, allowing multiple jets to be formed and a higher production rate to be achieved. Emulsion electrospinning increases the loading capacity of drug-polymer systems with low compatibility or affinity, such as water-soluble drugs or proteins loaded in a hydrophobic polymer for longer release. When compared with traditional blending methods, emulsion eliminates the need for a common solvent for both the drug and the polymer. Emulsifiers such as surfactants are frequently used to encapsulate and stabilize the drug phase [24, 71–76].

### 4.5 Solution electrospinning

Solution electrospinning is the most commonly used method wherein the polymer to be electrospun is dissolved in an appropriate solvent at a suitable concentration. As already discussed, the solution viscosity, as well as other conditions, needs to be optimized carefully to obtain nanofibers with desirable properties [5, 30, 31].

## 5. Applications of electrospun fibers

Electrospinning is employed extensively in industrial applications because of the attractive properties of electrospun fibers. The unique properties of electrospun fibers may be summarized as follows. First, electrospun fibers have diameters ranging from micro to nanometers. Second, the fibers are porous and aligned. Third, the electrospun fibers have a large aspect ratio and a high surface-to-volume ratio. Fourth, electrospinning enables the production of fibers with an infinite number of chemical compositions, and fifth, it also allows the production of different types of morphology by modifying the spinneret. With the combination of these properties, electrospun fibers can be utilized in biomedical applications [12, 41, 43, 51, 54, 60, 77–84], filtration [39, 85–88], energy sectors [2, 89–93], sensors [5, 26, 94–97], textiles, catalysis [26, 98], and electrical applications [99, 100]. **Figure 7** shows the different types of nanofibers and their applications.

### 5.1 Biomedical applications

In biomedical applications, biocompatible polymers with bioadhesive and biodegradable properties are preferred. Material selection is critical in the production of these nanofibers because it affects their morphology, biocompatibility, mechanical

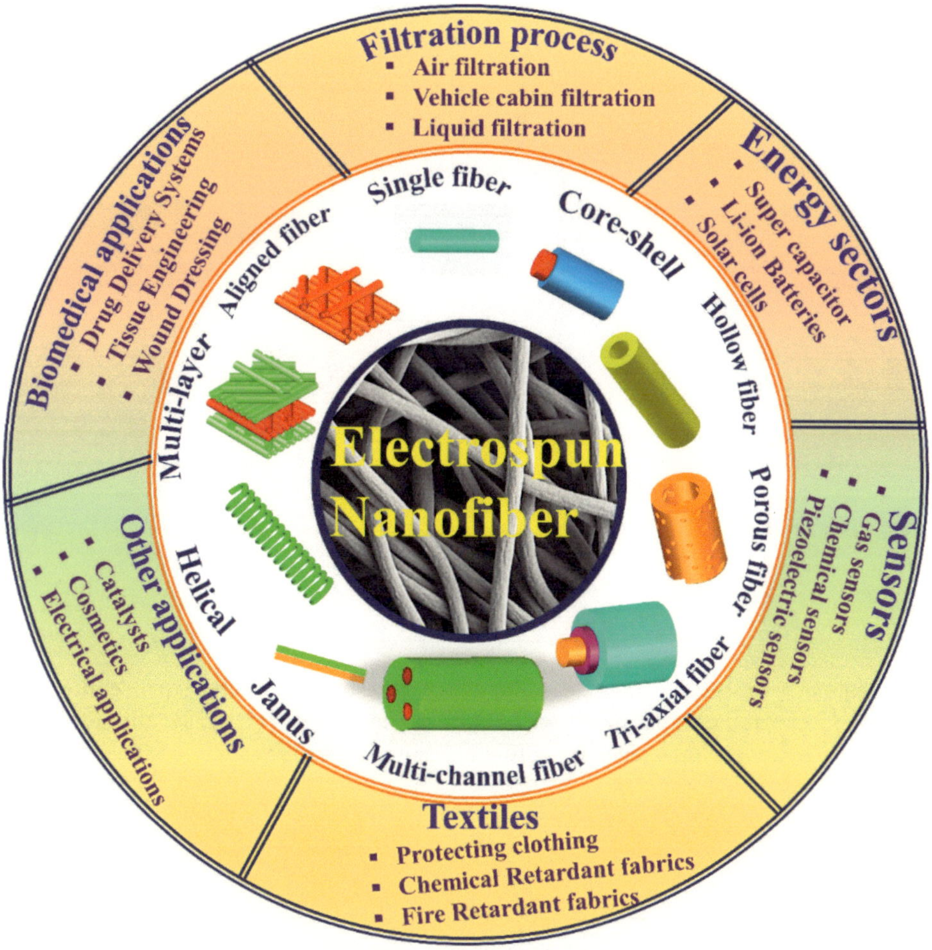


**Figure 7.**
*Different types of electrospun nanofibers and their applications.*

strength, degradation rate, and release profile, as well as their interactions with cells, which can result in a range of tissue responses.

Jalaja et al. [43] fabricated electrospun core-shell structured gelatin-chitosan nanofibers for biomedical applications. Chitosan as the shell can mimic the extracellular matrix while gelatin in the core can incorporate drugs and bioactive molecules. Singh et al. [54] fabricated core-shell nanofibers for biomedical applications using a novel coaxial airbrushing method as shown in **Figure 8**.

The core-shell nanofiber was fabricated using an air brush with a coaxial needle to flow two distinct polymeric solutions containing biomolecules [polyethylene oxide (PEO)/poly-DL-lactide/PCL (polycaprolactone)] in core and PCL/PEO is in shell. The great potential of coaxial electrospinning in biomedical field is evidenced by a large number of reports and reviews available in the literature on the topic [54].

### *5.1.1 Drug delivery systems*

The single nozzle electrospinning produces fibers with a high surface area to volume ratio, and as a result, more amounts of drugs would be present at the fiber surface.

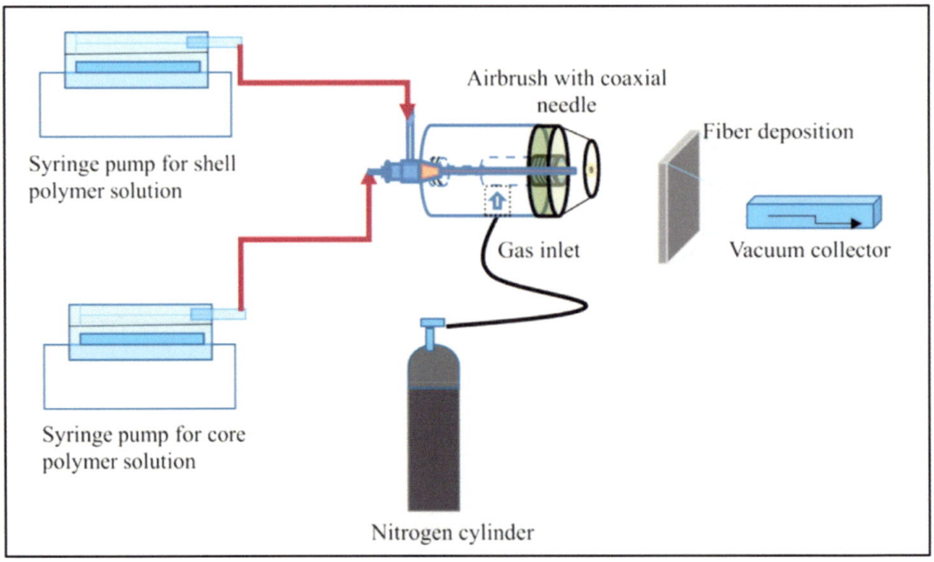


**Figure 8.**
*Experimental setup for core-shell fiber preparation via coaxial airbrushing methods [54].*

This frequently leads to a burst release, in which a large amount of the drug content is released into the solution quickly at the start of the process [11, 47, 84, 101, 102]. So, many researchers have relied on core-shell nanofibers to regulate the rate of drug release [34, 42, 45, 49, 103].

### *5.1.2 Tissue engineering*

The interaction of cells and scaffolds for the secretion of extracellular matrix (ECM) and the development of new tissues are the core ingredients of tissue engineering. Biocomposite scaffolds were created to overcome restrictions such as inflammation, toxicity, and recognition caused by scaffold degradation. In tissue engineering, electrospun nanofiber scaffolds are used to replicate the function of the native extracellular matrix (ECM). Core-shell is a flexible approach with intriguing features for encapsulation of the active component, such as growth factors and drugs in the core of the fiber, which is advantageous in drug delivery and tissue regeneration [42, 48, 49, 53, 73, 81, 104–109].

### *5.1.3 Wound healing*

Wound healing is a complicated and comprehensive process that includes four phases: hemostasis, inflammation, proliferation, and remodeling. These stages involve a coordinated and integrated process including extracellular matrix (ECM) mediators, cell growth factors, platelets, cytokines, and chemokines, among others. Ideal wound healing dressings are planned to be multifunctional and capable of delivering various drugs required at various phases of healing. There are various advantages of using electrospun nanofiber for wound dressing. First, the morphology and microstructures of electrospun nanofibers are similar to the natural ECM, which offers a perfect microenvironment for cell adhesion, proliferation, migration, and differentiation. Secondly, the electrospun nanofibers can simultaneously combine the biocompatibility of natural polymers and the reliable mechanical strength of synthetic polymers. The rate and timing of drug release may be controlled by changing the fiber structure to promote effective wound healing. As a result, electrospun nanofibers have a lot of potential for developing improved bioactive wound dressings [26, 41, 47, 49, 53, 54, 66, 77, 78, 98, 108, 110, 111].

### 5.2 Applications in filtration process

Electrospun fibers are suitable for filtration due to the properties, particularly, controlled porosity and high surface area to volume ratio. The potential of electrospun fibers in air filtration process has been well investigated [39, 87, 88, 112, 113].

Controlling the porosity of the nanofiber mesh is crucial for air particle permeability and avoiding hazardous particulate matter including dust, pollen, and bacteria. The electrospun fibers possess layer-by-layer structure that provides many interaction sites for a high number of airborne particles. Because of their characteristics, electrospun nanofibers can be used for air filtration [4, 19, 31, 114]. Zhang et al. [14] fabricated electrospun ultrafine fibers for advanced face masks with capability to physically block viruses.

### 5.3 Applications in energy sectors

The one-dimensional electrospun nanofibers have nanostructured materials with a high aspect ratio, strong mechanical strength, and efficient electron transport, allowing for a wide range of applications in the energy sectors. Additional advantages of electrospun nanofibers include: (1) attractive properties (unidirectional electron flow, uniform porosity, excellent aspect ratio, high surface area, tunable wettability, fine flexibility, and high connectivity); (2) easily tunable morphologies and characteristics according to the precursor solution, processing settings, and setup geometries; (3) simple preparation procedures at the lab scale and feasibility to be made on an industrial basis from a variety of materials; (4) capability to act as free standing electrodes without the need of conductive agents and binders; (5) possibilities to further improve the properties with simple post-electrospinning procedures (solvothermal method, calcination, electrodeposition, chemical vapor deposition, etc.). As a result, electrospun 1D nanofibers have shown significant promise as suitable materials for supercapacitors [6, 30, 92, 115–117], Li-ion batteries [2, 7, 25, 29, 118–123], solar cells [2, 90, 124–128], etc.

#### *5.3.1 Supercapacitors*

Electrospun nanofibers are incredibly versatile in terms of forming unique structures that may be altered with defects, functional groups, and other active materials, which is crucial for overcoming the present challenges toward developing efficient supercapacitors. Several research articles and review papers have discussed the attempts to exploit the potential of electrospun nanofibers as components of supercapacitors [2, 6, 22, 30, 89, 92, 115, 117].

#### *5.3.2 Li-ion batteries (LIBs)*

LIBs have sparked academic and industry interest because of their long cycle life, high energy density, high operational voltage, and low self-discharge rate.

Many advancements in LIB technology would not have been possible without the development of nanocomposites and nanometer-thick coatings to increase ionic and electronic conduction channels and prevent undesirable and irreversible side reactions. Electrospun nanofibers have excellent electrical and ionic conductivity due to their tunable fiber diameter, high porosity, high specific surface area, and interconnected pore structure, which is useful for improving cyclability and rate capability. Electrospun carbon nanofiber anodes loaded with metal oxide have sparked attention as anode materials in LIBs due to their high theoretical capacity, longer cycle life, and quick recharging rates [2, 118–122, 129–132].

### *5.3.3 Solar cells*

Many inorganic precursors may be electrospun with the help of carrier polymers and then annealed to produce inorganic fibers. This has sparked interest in dye-sensitized solar cells (DSSCs) and other solar energy-generating technologies [2, 124–128, 133, 134]. Nanofibers of $TiO_2$ have been produced, which are typically employed as the photo-sensitized anode in DSSCs. Nagata et al. [134] utilized coaxial electrospinning to fabricate solar cells using heterojunction polymer fibers.

## 5.4 Sensors

Sensors have long been a prominent research topic among all the applications of nanomaterials. A sensor is a device that can be detected, measured, and converted into meaningful output signals by following certain rules. In diverse sectors, sensors are also known as sensitive components, detectors, converters,

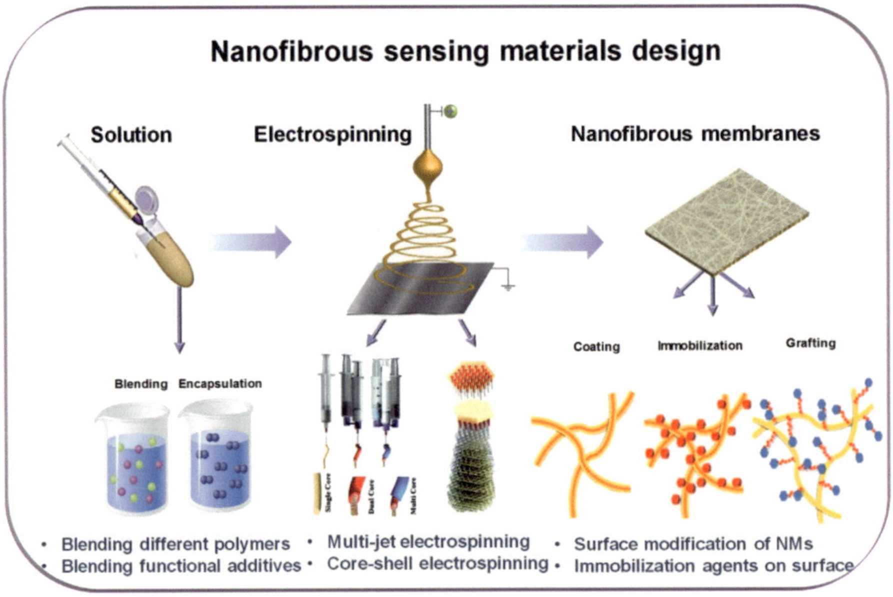


**Figure 9.**
*Schematic illustration of the fabrication of sensing materials via electrospinning [5].*

and so on [5, 26, 95–97]. Researchers have designed a number of unique methods for fabricating nanomaterials using conventional electrospinning to satisfy the sensor response unit's specifications as shown in **Figure 9** [5]. Many electrospun nanofiber-based nanomaterials are developed as gas sensors, chemical sensors, piezoelectric sensors [135, 136], and biosensors [137–140].

### 5.5 Textiles and other applications

Electrospun nanofibers are found to be potential alternative reinforcing choice for composites due to their superior mechanical as well as tunable chemical and physical properties. Despite the fact that electrospun nanofibers offer a number of desired features for usage as nano-fillers, little investigation into their application as prospective reinforcements has been done. The restriction is most likely due to their specific type of nano-fibrous architecture, which they may be made with. The electrospinning technology has been used to incorporate nanoparticles into composite nanofiber fabrics to impart new functions. Additional elements, such as metal nanoparticles, metal oxide nanoparticles, ionic liquids, and conductive polymers, might be included in these nano-fibrous structures to provide technical functions to an engineered fabric, such as electrical conductivity, strain resistance, and antibacterial qualities. The most often used nanomaterials are silver (Ag) and titanium dioxide ($TiO_2$), followed by silicon dioxide ($SiO_2$), carbon nanotubes (CNTs), and zinc oxide (ZnO), although a number of polymers are being studied for a range of prospective purposes. In textile applications, polymers such as polyester, polyamide, polyacrylonitrile (PAN), and polyethylene oxide (PEO) are often used. The composited nanofiber textiles with polymer and nanomaterial filler are expected to be light, strong, mechanically flexible, cost-effective, and simple to manufacture. In the case of food science applications, however, only food-grade polymers and safe, nontoxic solvents should be used to make nanofibers. As a result, natural polymers including collagen, gelatin, elastin, fibrinogen, and chitosan have been employed in electrospun technologies in a variety of applications [43, 77, 141]. Low basic weight, small fiber diameter, pore size, high surface area, and fiber chemistry are all critical in the selection of materials for specific applications. With better quality control of the electrospinning process, the fabrication of nanofiber-based textiles from electrospun nanofiber-based material might be widely commercialized [13, 79, 98, 142–144].

## 6. Summary and conclusion

Electrospinning allows for infinite combinations of chemical compositions, as defined by the periodic table, with morphologies and structures that may be controlled using solution and process parameters. This opens up a vast array of one-dimensional nanomaterials with various desired characteristics that might be used in a wide range of applications. This chapter explained the basic principles of electrospinning and information on how to carry it out. The history of electrospinning and principle, the configuration of electrospinning setup, parameters that influence the properties of electrospun fiber, various types of spinnerets to fabricate electrospun fibers, and their applications in diverse fields have been discussed in brief.

## 7. Future prospects

Significant development has been made in the field of electrospinning in the previous decade than ever before, and technical breakthroughs continue to evolve. Despite substantial advances, there are issues to be addressed in the future.

It's still difficult to synthesize uniform electrospun nanofibers with certain morphological, mechanical, and chemical properties that are tailored to specific end-user applications. Furthermore, a better understanding is needed to intelligently regulate the processing conditions and solution parameters to impact fiber mat characteristics for specialized applications. Natural polymers have poor chemical and mechanical qualities in some circumstances, and they are less widely employed in various applications than synthetic polymers. As a result, new hybrid polymer systems based on synthetic and natural polymers that are electrospinnable and have better functionalities are needed for a wide range of applications, particularly in biotechnology.

Several spinnerets have been proposed for mass production; however, some of them may not be suitable, and some of them still require additional experimental verifications in terms of fiber quality control and the electrospinning process. Electrospinning processes such as coaxial, tri-axial, multichannel, and side-by-side electrospinning have been found to be effective in customizing the physical properties of electrospun nanofibers. However, more research into the process control and mechanism for core-shell and multichannel morphologies of electrospun nanofibers is required. In comparison to solution electrospinning, melt electrospinning provides benefits such as the lack of harmful solvents and high production. To achieve theoretically expected strengths, however, tremendous measures are needed to minimize fiber diameters while simultaneously establishing a high degree of orientation in the structure. Needleless spinnerets offer considerable promise in electrospinning nanofibers on large scales, but needleless electrospinning of bicomponent nanofibers remains a difficulty. Although over 200 polymers have been successfully electrospun into nanofibers, there has been little research on the macromolecular orientation and crystalline structures of the resultant fibers. In addition, the influence of the crystalline phase and molecule orientation on mechanical properties remains unclear.

Application of electrospun nanofibers: Despite the fact that electrospun nanofibers have been shown to be a potential candidate for composite reinforcements, little study has been done on this subject due to their poor mechanical properties when compared with standard alternatives such as carbon or glass fibers. Adding nanoparticles as a post-treatment to electrospun nanofibers may be an acceptable approach to increase mechanical properties. One of the main hurdles to the advancement of electrospinning applications in tissue engineering is increasing scaffold thickness, pore size, and the difficulty to prepare identical scaffolds. The relationships between drug-controlled release profiles and nanofiber structures should be investigated in detail. To understand the drug transport mechanism and to predict drug release kinetics as a function of nanofiber structure, mathematical models of drug release from diverse nanofibers might be useful. Electrospun nanofibers offer a lot of potential in the energy sector, but there are still a lot of obstacles and opportunities to be explored in this field. "Bottleneck" difficulties in electrospun nanofibers such as poor conductivity, chemical structural instability, volume expansion, and kinetic hysteresis of active materials should be addressed in this research field. For filtering applications, further research might involve using different polymers and/or post-treatment procedures to control the formation of pores on fibers, as well as a better understanding of the transport process through fibers with microporous rough surfaces. Electrospun nanofibers have been shown to exhibit remarkable properties for sensor applications. Further research into improving the surface area and pore sizes of nano-fibrous membranes is suggested to improve sensitivity. Furthermore, incorporating such nanomaterials into practical devices is difficult, as it necessitates materials with precise orientation, size, and repeatability in order to place them in specific positions and orientations.

## Acknowledgements

The authors thank Director, IIST, for the support and the facilities provided. Govind Kumar Sharma thanks University Grants Commission, New Delhi, India, for a Research Fellowship.

## Conflict of interest

The authors declare that they do not have any conflict of interest.

## Author details

Govind Kumar Sharma and Nirmala Rachel James*
Department of Chemistry, Indian Institute of space science and Technology, Thiruvanthanumpuram, Kerala, India

*Address all correspondence to: nirmala@iist.ac.in

## References

[1] Zhang Z, Ji D, He H, Ramakrishna S. Electrospun ultrafine fibers for advanced face masks. Materials Science and Engineering R: Reports. 2020;**2021**(143):100594. DOI: 10.1016/j.mser.2020.100594

[2] Shi X, Zhou W, Ma D, Ma Q, Bridges D, Ma Y, et al. Electrospinning of nanofibers and their applications for energy devices. Journal of Nanomaterials. 2015;**2015**:140716. DOI: 10.1155/2015/140716

[3] Williams GR, Raimi-Abraham BT, Luo CJ. Coaxial and multi-axial electrospinning. In: Nanofibres in Drug Delivery. London: UCL Press; 2018. pp. 106-148. Available from: http://www.jstor.org/stable/j.ctv550dd1.8

[4] Xue J, Wu T, Dai Y, Xia Y. Electrospinning and electrospun nanofibers: Methods, materials, and applications. Chemical Reviews. 2019;**119**(8):5298-5415

[5] Li Y, Abedalwafa MA, Tang L, Li D, Wang L. Electrospun nanofibers for sensors. In: Andrew W, editor. Electrospinning: Nanofabrication and Applications. New York, United States: Elsevier; 2018. pp. 571-601

[6] Bhattacharya S, Roy I, Tice A, Chapman C, Udangawa R, Chakrapani V, et al. High-conductivity and high-capacitance electrospun fibers for Supercapacitor applications. ACS Applied Materials & Interfaces. 2020;**12**(17):19369-19376

[7] Li Y, Li Q, Tan Z. A review of electrospun nanofiber-based separators for rechargeable lithium-ion batteries. Journal of Power Sources. 2019;**443**:227262

[8] Knight DP, Vollrath F. Liquid crystalline spinning of spider silk. Nature. 2001;**410**(6828):541-548. DOI: 10.1038/35069000

[9] Andersson M, Jia Q, Abella A, Lee XY, Landreh M, Purhonen P, et al. Biomimetic spinning of artificial spider silk from a chimeric minispidroin. Nature Chemical Biology. 2017;**13**(3):262-264

[10] Heim M, Keerl D, Scheibel T. Spider silk: From soluble protein to extraordinary fiber. Angewandte Chemie. 2009;**48**(20):3584-3596

[11] Yu DG, Li XY, Wang X, Yang JH, Bligh SWA, Williams GR. Nanofibers fabricated using Triaxial electrospinning as zero order drug delivery systems. ACS Applied Materials & Interfaces. 2015;7(33):18891-18897

[12] Davoodi P, Gill EL, Wang W, Shery Huang YY. Advances and innovations in electrospinning technology. In: Biomed Applications of Electrospinning and Electrospraying. Sawston, Cambridge: Woodhead Publishing, Elsevier; 2021. pp. 45-81

[13] Akdere M, Schneiders T. Modeling of the electrospinning process. In: Advances in Modeling and Simulation in Textile Engineering. Sawston, Cambridge: Woodhead Publishing, Elsevier; 2021. pp. 237-253

[14] Zhang Z, Ji D, He H, Ramakrishna S. Electrospun ultrafine fibers for advanced face masks. Materials Science and Engineering R: Reports. 2021;**143**:100594

[15] Hou J, Yang Y, Yu DG, Chen Z, Wang K, Liu Y, et al. Multifunctional fabrics finished using electrosprayed hybrid Janus particles containing

nanocatalysts. Chemical Engineering Journal. 2021;**411**:128474

[16] Neumann C. Production, properties, and uses of the finest threads. Proceedings of the Physical Society of London. 1887;**489**:489-499

[17] Cooley JF. Apparatus for electrically dispersing fluids. US Pat 692,631. 1900;**693**(631):1-6 Available from: http://www.google.com/patents/US692631

[18] Soderlund H, Kaariainen L, Von Bonsdorff CH. Properties of Semliki Forest virus nucleocapsid. Medical Biology. 1975;**53**(5):412-417

[19] Xue J, Wu T, Dai Y, Xia Y. Electrospinning and electrospun nanofibers: Methods, materials, and applications. Chemical Reviews. 2019;**119**(8):5298-5415

[20] Yarin AL, Koombhongse S, Reneker DH. Bending instability in electrospinning of nanofibers. Journal of Applied Physics. 2001;**89**:3018. DOI: 10.1063/1.1333035

[21] Reneker DH, Yarin AL, Fong H, Koombhongse S. Bending instability of electrically charged liquid jets of polymer solutions in electrospinning. Journal of Applied Physics. 2000;**87**(9I):4531-4547

[22] Joshi B, Samuel E, Kim Y, Yarin AL, Swihart MT, Yoon SS. Review of recent progress in electrospinning-derived freestanding and binder-free electrodes for supercapacitors. Coordination Chemistry Reviews. 2022;**460**:214466

[23] Luo CJ, Stoyanov SD, Stride E, Pelan E, Edirisinghe M. Electrospinning versus fibre production methods: From specifics to technological convergence. Chemical Society Reviews. 2012;**41**:4708-4735 Available from: www.rsc.org/csr

[24] Yarin AL. Coaxial electrospinning and emulsion electrospinning of core-shell fibers. Polymers for Advanced Technologies. 2011;**22**(3):310-317

[25] Li Y, Li Q, Tan Z. A review of electrospun nanofiber-based separators for rechargeable lithium-ion batteries. Journal of Power Sources. 2019;**443**(August):227262. DOI: 10.1016/j.jpowsour.2019.227262

[26] Fadil F, Affandi NDN, Misnon MI, Bonnia NN, Harun AM, Alam MK. Review on electrospun nanofiber-applied products. Polymers (Basel). 2021;**13**(13):1-31

[27] Angel N, Li S, Yan F, Kong L. Recent advances in electrospinning of nanofibers from bio-based carbohydrate polymers and their applications. Trends in Food Science and Technology. 2022;**120**:308-324

[28] Kumar Sharma G, Rachel JN. Progress in electrospun polymer composite fibers for microwave absorption and electromagnetic interference shielding. ACS Applied Electronic Materials. 2021;**3**(11):4657-4680

[29] Lee BS. Polymers A review of recent advancements in electrospun anode materials to improve rechargeable lithium battery performance. Polymers. 2020;**12**(9):2035

[30] Liang J, Zhao H, Yue L, Fan G, Li T, Lu S, et al. Recent advances in electrospun nanofibers for supercapacitors. Journal of Materials Chemitry A. 2020;**8**:16747-16789

[31] Mamun A, Blachowicz T, Sabantina L. Electrospun nanofiber mats for filtering applications technology, structure and materials. Polymers (Basel). 2021;**13**(9):1-14

[32] Zhao K, Wang W, Yang Y, Wang K, Yu DG. From Taylor cone to solid nanofiber in tri-axial electrospinning: Size relationships. Results in Physics. 2019;**15**:102770. DOI: 10.1016/j.rinp.2019.102770

[33] Zargham S, Bazgir S, Tavakoli A, Rashidi AS, Damerchely R. The effect of flow rate on morphology and deposition area of electrospun nylon 6 nanofiber. Journal of Engineered Fibers and Fabrics. 2012;**7**(4):42-49

[34] Monfared M, Taghizadeh S, Zare-Hoseinabadi A, Mousavi SM, Hashemi SA, Ranjbar S, et al. Emerging frontiers in drug release control by core–shell nanofibers: A review. Drug Metabolism Reviews. 2019;**51**(4):589-611. DOI: 10.1080/03602532.2019.1642912

[35] Sharma GK, James NR. Carbon black incorporated carbon nanofiber based polydimethylsiloxane composite for electromagnetic interference shielding. Carbon Trends. 2022;**8**:100177

[36] Yang C, Jia Z, Xu Z, Wang K, Guan Z, Wang L. Comparisons of Fibers Properties between Vertical and Horizontal Type Electrospinning Systems. Annu Rep - Conf Electr Insul Dielectr Phenomena: CEIDP; 2009. pp. 204-207

[37] He XX, Zheng J, Yu GF, You MH, Yu M, Ning X, et al. Near-field electrospinning: Progress and applications. Journal of Physical Chemistry C. 2017;**121**(16):8663-8678

[38] Rathore P, Schiffman DJ. Beyond the single-nozzle: Coaxial electrospinning enables innovative nanofiber chemistries, geometries, and applications. ACS Applied Materials & Interfaces. 2020;**13**(1):48-66

[39] Sundarrajan S, Tan KL, Lim SH, Ramakrishna S. Electrospun nanofibers for air filtration applications. Procedia Engineering. 2014;**75**:159-163. DOI: 10.1016/j.proeng.2013.11.034

[40] Tang, Ben Zhong, Alaa S. Abd-El-Aziz SC. Electrospinning Series Editors: Titles in the Series. London: RSC Polymer Chemistry Series No.14; 2014

[41] Moradipour P, Limoee M, Janfaza S, Behbood L. Core-Shell nanofibers based on polycaprolactone/polyvinyl alcohol and polycaprolactone/collagen for biomedical applications. Journal of Pharmaceutical Innovation. 2021:1-10. DOI: 10.1007/s12247-021-09568-z

[42] Perez RA, Kim HW. Core-shell designed scaffolds for drug delivery and tissue engineering. Acta Biomaterialia. 2015;**21**(March):2-19. DOI: 10.1016/j.actbio.2015.03.013

[43] Jalaja K, Naskar D, Kundu SC, James NR. Potential of electrospun core–shell structured gelatin–chitosan nanofibers for biomedical applications. Carbohydrate Polymers. 2016;**136**:1098-1107

[44] Tong Z, Liao Z, Liu Y, Ma M, Bi Y, Huang W, et al. Hierarchical Fe3O4/Fe@C@MoS2 core-shell nanofibers for efficient microwave absorption. Carbon. 2021;**179**:646-654

[45] Zupančič Š. Core-shell nanofibers as drug delivery systems. Acta Pharmaceutica. 2019;**69**(2):131-153

[46] Neisiany RE, Saied Khorasani N, Kong J, Lee Y, Javad R, et al. Core-shell nanofibers for developing self-healing materials: Recent progress and future directions. Material Design & Processing Communications. 2021;**3**:90. DOI: 10.1002/mdp2.90

[47] Kharaghani D, Gitigard P, Ohtani H, Kim KO, Ullah S, Saito Y, et al. Design and characterization of dual drug delivery based on in-situ assembled PVA/PAN core-shell nanofibers for

wound dressing application. Scientific Reports. 2019;**9**(1):1-11. DOI: 10.1038/s41598-019-49132-x

[48] Ezhilarasu H, Sadiq A, Ratheesh G, Sridhar S, Ramakrishna S, Mohd MH, et al. Functionalized core/shell nanofibers for the differentiation of mesenchymal stem cells for vascular tissue engineering. Nanomedicine. 2019;**14**(2):201-214

[49] Li Z, Mei S, Dong Y, She F, Li P, Li Y, et al. Multi-Functional Core-Shell Nanofibers for Wound Healing. Nanomaterials. 2021;**11**(6):1546. DOI: 10.3390/nano11061546

[50] Lee S, Park J, Kim MC, Kim M, Park P, Yoon IJ, et al. Polyvinylidene fluoride Core-Shell nanofiber membranes with highly conductive shells for electromagnetic interference shielding. ACS Applied Materials & Interfaces. 2021;**13**(21):25428-25437

[51] Elahi MF, Lu W. Core-shell fibers for biomedical applications-A review. Journal of oJ Bioengineering & Biomedical Science. 2013;**03**(01):121

[52] Zhu LF, Zheng Y, Fan J, Yao Y, Ahmad Z, Chang MW. A novel core-shell nanofiber drug delivery system intended for the synergistic treatment of melanoma. European Journal of Pharmaceutical Sciences. 2019;**137**(June):105002. DOI: 10.1016/j.ejps.2019.105002

[53] Movahedi M, Asefnejad A, Rafienia M, Khorasani MT. Potential of novel electrospun core-shell structured polyurethane/starch (hyaluronic acid) nanofibers for skin tissue engineering: In vitro and in vivo evaluation. International Journal of Biological Macromolecules. 2020;**146**:627-637. DOI: 10.1016/j.ijbiomac.2019.11.233

[54] Singh R, Ahmed F, Polley P, Giri J. Fabrication and characterization of Core-Shell nanofibers using a next-generation airbrush for biomedical applications. ACS Applied Materials & Interfaces. 2018;**10**(49):41924-41934

[55] Ghosal K, Augustine R, Zaszczynska A, Barman M, Jain A, Hasan A, et al. Novel drug delivery systems based on triaxial electrospinning based nanofibers. Reactive and Functional Polymers 2021;**163** (December 2020):104895. DOI:10.1016/j.reactfunctpolym.2021.104895

[56] Nagiah N, Murdock CJ, Bhattacharjee M, Nair L, Laurencin CT. Development of Tripolymeric Triaxial electrospun fibrous matrices for dual drug delivery applications. Scientific Reports. 2020;**10**(1):1-11

[57] Han D, Steckl AJ. Triaxial electrospun nanofiber membranes for controlled dual release of functional molecules. ACS Applied Materials & Interfaces. 2013;**5**(16):8241-8245

[58] Lallave M, Bedia J, Ruiz-Rosas R, Rodríguez-Mirasol J, Cordero T, Otero JC, et al. Filled and hollow carbon nanofibers by coaxial electrospinning of alcell lignin without binder polymers**. Advanced Materials. 2007;**19**:4292-4296

[59] Zhao Y, Cao X, Jiang L. Bio-mimic multichannel microtubes by a facile method. Journal of the American Chemical Society. 2007;**129**(4):764-765

[60] Partheniadis I, Nikolakakis I, Laidmäe I, Heinämäki J. A mini-review: Needleless electrospinning of nanofibers for pharmaceutical and biomedical applications. 2020;**8**(6):673

[61] Iranshahi K, Schoeller J, Luisier N, Chung M, Hashemizadeh S, Fortunato G, et al. Improving needleless electrospinning throughput by tailoring polyurethane solution properties with

Polysiloxane additives. ACS Applied Polymer Materials. 2022;**4**(3):2205-2215

[62] Hwang M, O. Karenson M, A. Elabd Y. High production rate of high purity, high Fidelity Nafion nanofibers via needleless electrospinning. ACS Applied Polymer Materials. 2019;**1**(10):2731-2740

[63] Yang Z, Zhang X, Qin Z, Li H, Wang J, Zeng G, et al. Airflow synergistic needleless electrospinning of instant noodle-like curly nanofibrous membranes for high-efficiency air filtration. Small. 2022;**18**:2107250

[64] Cao L, Su D, Su Z, Chen X. Fabrication of multiwalled carbon nanotube/polypropylene conductive fibrous membranes by melt electrospinning. Industrial and Engineering Chemistry Research. 2014;**53**(6):2308-2317

[65] Li X, Zheng Y, Mu X, Xin B, Lin L. Investigation into jet motion and fiber properties induced by electric fields in melt electrospinning. Industrial & Engineering Chemistry Research. 2020;**59**(5):2163-2170

[66] Shao Z, Chen J, Ke L jie, Wang Q, Wang X, Li W, et al. Directional transportation in a self-pumping dressing based on a melt electrospinning hydrophobic mesh. ACS Biomaterials Science & Engineering. 2021;**7**(12):5918-5926

[67] Shabani E, Azimi Yancheshme A, Ronen A, E. Gorga R. Effect of the spin-line temperature profile on the translocation of the solidification point and jet thinning in unconfined melt electrospinning. ACS Applied Polymer Materials. 2020;**3**(1):268-278

[68] Chen F, Hochleitner G, Woodfield T, Groll J, D. Dalton P, G. Amsden B. Additive manufacturing of a photo-cross-linkable polymer via direct melt electrospinning writing for producing high strength structures. Biomacromolecules. 2015;**17**(1):208-214

[69] Xie G, Wang Y, Han X, Gong Y, Wang J, Zhang J, et al. Pulsed electric fields on poly-l-(lactic acid) melt electrospun fibers. Industrial & Engineering Chemistry Research. 2016;**55**(26):7116-7123

[70] Dalton DP, Klinkhammer K, Salber J, Klee D, Möller M. Direct in vitro electrospinning with polymer melts. Biomacromolecules. 2006;**7**(3):686-690

[71] Sharif F, Arjmand M, Moud AA, Sundararaj U, Roberts EPL. Segregated hybrid poly(methyl methacrylate)/graphene/magnetite nanocomposites for electromagnetic interference shielding. ACS Applied Materials & Interfaces. 2017;**9**(16):14171-14179

[72] Zhang C, Zhang H. Formation and stability of Core–Shell nanofibers by electrospinning of gel-like corn oil-in-water emulsions stabilized by gelatin. Journal of Agricultural and Food Chemistry. 2018;**66**(44):11681-11690

[73] Pal P, Kumar Srivas P, Dadhich P, Das B, Maulik D, Dhara S. Nano–/microfibrous cotton-wool-like 3D scaffold with Core–Shell architecture by emulsion electrospinning for skin tissue regeneration. ACS Biomaterials Science & Engineering. 2017;**3**(12):3563-3575

[74] Choi SH, Youn DY, Mu Jo S, Oh SG, Kim ID. Micelle-mediated synthesis of single-crystalline β(3C)-SiC fibers via emulsion electrospinning. ACS Applied Materials & Interfaces. 2011;**3**(5):1385-1389

[75] Zhao P, Soin N, Prashanthi K, Chen J, Dong S, Zhou E, et al. Emulsion electrospinning of polytetrafluoroethylene (PTFE) Nanofibrous membranes

for high-performance triboelectric Nanogenerators. ACS Applied Materials & Interfaces. 2018;**10**(6):5880-5891

[76] Qi H, Hu P, Xu J, Wang A. Encapsulation of drug reservoirs in fibers by emulsion electrospinning: Morphology characterization and preliminary release assessment. Biomacromolecules. 2006;**7**(8):2327-2330

[77] Hoque ME, Nuge T, Yeow TK, Nordin N. Electrospun matrices from natural polymers for skin regeneration. In: Nanostructured Polymer Composites for Biomedical Applications. Amsterdam, Netherlands: Elsevier; 2019. pp. 87-104

[78] Hoque ME, Nuge T, Yeow TK, Nordin N. Electrospun matrices from natural polymers for skin regeneration. In: Nanostructured Polymer Composites for Biomedical Applications. Amsterdam, Netherlands: Elsevier; 2019. pp. 87-104

[79] Joseph J, Nair VS, Menon D. Integrating Substrateless electrospinning with textile Technology for Creating Biodegradable Three-Dimensional Structures. Nano Letters. 2015;**15**(8):5420-5426

[80] Sagitha P, Reshmi CR, Manaf O, Sundaran SP, Juraij K, Sujith A. Development of nanocomposite membranes by electrospun nanofibrous materials. In: Nanocomposite Membranes for Water and Gas Separation. Amsterdam, Netherlands: Elsevier; 2020. pp. 199-218

[81] Sarika PR, Cinthya K, Jayakrishnan A, Anilkumar PR, James NR. Modified gum arabic cross-linked gelatin scaffold for biomedical applications. Materials Science and Engineering: C. 2014;**43**:272-279

[82] Elahi MF, Lu W. Core-shell fibers for biomedical applications-A review. Journal of Bioengineering and Biomedical Science. 2013;**03**(01):1-14

[83] Jalaja K, Kumar PRA, Dey T, Kundu SC, James NR. Modified dextran cross-linked electrospun gelatin nanofibres for biomedical applications. Carbohydrate Polymers. 2014;**114**:467-475

[84] Davoodi P, Gill EL, Wang W, Shery Huang YY. Advances and innovations in electrospinning technology. In: Biomedical Applications of Electrospinning and Electrospraying. Sawston, Cambridge: Woodhead Publishing, Elsevier; 2021. pp. 45-81

[85] Jiang S, Hou H, Agarwal S, Greiner A. Polyimide nanofibers by "green" electrospinning via aqueous solution for filtration applications. ACS Sustainable Chemistry & Engineering. 2016;**4**(9):4797-4804

[86] Graham K, Ouyang M, Raether T, Grafe T, Mcdonald B, Knauf P. Polymeric nanofibers in air filtration applications. Galveston, Texas: Fifteenth Annual Technical Conference & Expo of the American Filtration & Separations Society; April 9-12, 2002

[87] Zhu M, Han J, Wang F, Shao W, Xiong R, Zhang Q, et al. Electrospun nanofibers membranes for effective air filtration. Macromolecular Materials and Engineering. 2017;**302**(1):1-27

[88] Choi J, Joon Yang B, Bae GN, Hee JJ. Herbal extract incorporated nanofiber fabricated by an electrospinning technique and its application to antimicrobial air filtration. ACS Applied Materials & Interfaces. 2015;**7**(45):25313-25320

[89] Bonso JS, Kalaw GD, Ferraris JP. High surface area carbon nanofibers derived from electrospun PIM-1 for

energy storage applications †. Journal of Materials Chemistry A. 2014;**2**(2): 418-424. Available from: www.rsc.org/MaterialsA

[90] Bredar ARC, Chown AL, Burton AR, Farnum BH. Electrochemical impedance spectroscopy of metal oxide electrodes for energy applications. ACS Applied Energy Materials. 2020;**3**(1):66-98

[91] Sankar SS, Rathishkumar A, Geetha K, Kundu S. Electrospinning as a tool in fabricating hydrated porous cobalt phosphate fibrous network as high rate OER electrocatalysts in alkaline and neutral media. International Journal of Hydrogen Energy. 2021;**46**(17):10366-10376

[92] Shi K, P. Giapis K. Scalable fabrication of Supercapacitors by nozzle-free electrospinning. ACS Applied Energy Materials. 2018;**1**(2):296-300

[93] Chen R, Wang HY, Miao J, Yang H, Liu B. A flexible high-performance oxygen evolution electrode with three-dimensional NiCo2O4 core-shell nanowires. Nano Energy. 2015;**11**:333-340. DOI: 10.1016/j.nanoen.2014.11.021

[94] Shingange K, Swart HC, Mhlongo GH. Design of porous p-type LaCoO3 nanofibers with remarkable response and selectivity to ethanol at low operating temperature. Sensors and Actuators B: Chemical. 2020;**308**:127670

[95] Li W, Zhang LS, Wang Q, Yu Y, Chen Z, Cao CY, et al. Low-cost synthesis of graphitic carbon nanofibers as excellent room temperature sensors for explosive gases †. Journal of Materials Chemistry. 2012;**22**:15342. DOI: 10.1039/c2jm32031b

[96] Brosha EL, Mukundan R, Brown DR, Garzon FH, Visser JH, Zanini M, et al. CO/HC sensors based on thin films of LaCoO3 and La0.8Sr0.2CoO3−δ metal oxides. Sensors and Actuators B: Chemical. 2000;**69**(1-2):171-182

[97] Li W, Zhang LS, Wang Q, Yu Y, Chen Z, Cao CY, et al. Low-cost synthesis of graphitic carbon nanofibers as excellent room temperature sensors for explosive gases. Journal of Materials Chemistry. 2012;**22**(30):15342-15347

[98] Zhong W. Nanofibres for medical textiles. In: Advances in Smart Medical Textiles. Sawston, Cambridge: Woodhead Publishing, Elsevier; 2016. pp. 57-70

[99] Choi HJ, Kim MS, Ahn D, Yeo SY, Lee S. Electrical percolation threshold of carbon black in a polymer matrix and its application to antistatic fibre. Scientific Reports. 2019;**9**(1):1-12. DOI: 10.1038/s41598-019-42495-1

[100] Hou Y, Cheng L, Zhang Y, Du X, Zhao Y, Yang Z. High temperature electromagnetic interference shielding of lightweight and flexible ZrC/SiC nanofiber mats. Chemical Engineering Journal. 2021;**404**(2020):126521

[101] Bay Stie M, Corezzi MD, Juncos Bombin A, Ajalloueian F, Attrill E, Pagliara S, et al. Waterborne electrospinning of α-Lactalbumin generates tunable and biocompatible nanofibers for drug delivery. ACS Applied Nano Materials. 2020;**3**(2):1910-1921

[102] Moazeni N, Sadrjahani M. Stimuli-responsive nanofibrous materials in drug delivery systems. In: Engineered Polymeric Fibrous Materials. Sawston, Cambridge: Woodhead Publishing, Elsevier; 2021. pp. 171-189

[103] Pant B, Park M, Park SJ. Drug delivery applications of core-sheath nanofibers prepared by

coaxial electrospinning: A review. Pharmaceutics. 2019;**11**(7):305

[104] Matthews AJ, Wnek EG, Simpson GD, Bowlin LG. Electrospinning of collagen nanofibers. Biomacromolecules. 2002;**3**(2):232-238

[105] Lim DJ. Cross-linking agents for electrospinning-based bone tissue engineering. International Journal of Molecualr Sciences. 2022;**23**:5444. DOI: 10.3390/ijms23105444

[106] Nemati S, Jeong KS, Shin YM, Shin H. Current progress in application of polymeric nanofibers to tissue engineering. Nano Convergence. 2019;**6**(1):36. DOI: 10.1186/s40580-019-0209-y

[107] Rahmati M, Mills DK, Urbanska AM, Saeb MR, Venugopal JR, Ramakrishna S, et al. Electrospinning for tissue engineering applications. Progress in Materials Science. 2021;**117**:100721

[108] Wnek EG, Carr EM, Simpson GD, Bowlin LG. Electrospinning of nanofiber fibrinogen structures. Nano Letters. 2002;**3**(2):213-216

[109] Jalaja K, Sreehari VS, Kumar PRA, Nirmala RJ. Graphene oxide decorated electrospun gelatin nanofibers: Fabrication, properties and applications. Materials Science and Engineering: C. 2016;**64**:11-19

[110] Lv P, Xu W, Li D, Feng Q, Yao Y, Pang Z, et al. Metal-based bacterial cellulose of sandwich nanomaterials for anti-oxidation electromagnetic interference shielding. Materials and Design. 2016;**112**:374-382. DOI: 10.1016/j.matdes.2016.09.100

[111] Rezvani Ghomi E, Khosravi F, Neisiany RE, Shakiba M, Zare M, Lakshminarayanan R, et al. Advances in electrospinning of aligned nanofiber scaffolds used for wound dressings. Current Opinion in Biomedical Engineering. 2022;**22**:100393

[112] Kang L et al. Preparation of electrospun nanofiber membrane for air filtration and process optimization based on BP neural network. Materials Research Express. 2021;**8**:115010. DOI: 10.1088/2053-1591/ac37d6

[113] Satilmis B. Electrospinning polymers of intrinsic microporosity (PIMs) ultrafine fibers; preparations, applications and future perspectives. Current Opinion in Chemical Engineering. 2022;**36**:100793

[114] Deeraj BDS, Jayan JS, Saritha A, Joseph K. Electrospun biopolymer-based hybrid composites. In: Hybrid Natural Fiber Composites Material Formulations, Processing, Characterization, Properties and Engineering Applications. Sawston, Cambridge: Woodhead Publishing, Elsevier; 2021. pp. 225-252

[115] Dong Q, Wang G, Hu H, Yang J, Qian B, Ling Z, et al. Ultrasound-assisted preparation of electrospun carbon nanofiber/graphene composite electrode for supercapacitors. Journal of Power Sources. 2013;**243**:350-353. DOI: 10.1016/j.jpowsour.2013.06.060

[116] Bhattacharya S, Roy I, Tice A, Chapman C, Udangawa R, Chakrapani V, et al. High-conductivity and high-capacitance electrospun fibers for Supercapacitor applications. ACS Applied Materials & Interfaces. 2020;**12**(17):19369-19376

[117] Yang CH, Hsiao YC, Lin LY. Novel In situ synthesis of freestanding carbonized ZIF67/polymer nanofiber electrodes for Supercapacitors via electrospinning and pyrolysis techniques. ACS Applied Materials & Interfaces. 2021;**13**(35):41637-41648

[118] Zhai Y, Liu H, Li L, Yu J, Ding B. Electrospun Nanofibers for Lithium-Ion Batteries. Electrospinning: Nanofabrication and Applications. In: Andrew W, editor. Norwich, New York: Elsevier Inc.; 2018. pp. 671-694. DOI: 10.1016/B978-0-323-51270-1.00022-4

[119] Weng W, Kurihara R, Wang J, Shiratori S. Electrospun carbon nanofiber-based composites for lithium-ion batteries: Structure optimization towards high performance. Composites Communications. 2019;**15**:135-148

[120] Kannan SK, Hareendra-krishnakumar H, Joseph MG. Efficient polysulfide shuttle mitigation by graphene-lithium cobalt vanadate hybrid for advanced lithium-sulfur batteries. Journal of Electroanalytical Chemistry. 2021;**899**(June):115665. DOI: 10.1016/j.jelechem.2021.115665

[121] Di Carli M, Caso MF, Aurora A, Della SL, Rinaldi A, Ferrone E, et al. Electrospinning nanofibers as separators for lithium-ion batteries. AIP Conference Proceedings. 2019;**2145**(August)

[122] Liu G, Chen H, Xia L, Wang S, Ding LX, Li D, et al. Hierarchical mesoporous/macroporous perovskite La0.5Sr0.5CoO3-x nanotubes: A Bifunctional catalyst with enhanced activity and cycle stability for rechargeable lithium oxygen batteries. ACS Applied Materials & Interfaces. 2015;**7**(40):22478-22486

[123] Kang Y, Deng C, Liu X, Liang Z, Li T, Hu Q, et al. Binder-free electrode based on electrospun-fiber for Li ion batteries via a simple rolling formation. Nanoscale Research Letters. 2020;**15**(1):147

[124] Haghighat Bayan MA, Afshar Taromi F, Lanzi M, Pierini F. Enhanced efficiency in hollow core electrospun nanofiber-based organic solar cells. Scientific Reports. 2021;**11**(1):1-11. DOI: 10.1038/s41598-021-00580-4

[125] Fathy M, Kashyout AB, El Nady J, Ebrahim S, Soliman MB. Electrospun polymethylacrylate nanofibers membranes for quasi-solid-state dye sensitized solar cells. Alexandria Engineering Journal. 2016;**55**(2):1737-1743

[126] Jose Varghese R, Sakho EHM, Parani S, Thomas S, Oluwafemi OS, Wu J. Introduction to nanomaterials: Synthesis and applications. In: Nanomaterials for Solar Cell Applications. Amsterdam, Netherlands: Elsevier; 2019. pp. 75-95

[127] Zhu P, Sreekumaran Nair A, Shengjie P, Shengyuan Y, Ramakrishna S. Facile fabrication of TiO2–graphene composite with enhanced photovoltaic and photocatalytic properties by electrospinning. ACS Applied Materials & Interfaces. 2012;**4**(2):581-585

[128] Gerardo López-Covarrubias J, Soto-Muñoz L, Iglesias AL, Jesús V-GL. Electrospun nanofibers applied to dye solar sensitive cells: A review. Materials. 2019;**12**(19):3190

[129] Zhu LF, Zheng Y, Fan J, Yao Y, Ahmad Z, Chang MW. A novel core-shell nanofiber drug delivery system intended for the synergistic treatment of melanoma. European Journal of Pharmaceutical Sciences. 2019;**137**:105002

[130] Zhang Y, Zhang X, Ravi S, Silva P, Ding B, Zhang P, et al. Lithium-sulfur batteries meet electrospinning: Recent advances and the key parameters for high gravimetric and volume energy density. Advanced Science. 2021;**9**:2103879. DOI: 10.1002/advs.202103879

[131] Lei W, Li H, Tang Y, Huaiyu S, Shao H, Kong GH. Progress and

perspectivesonelectrospinningtechniques for solid-state lithium batteries. Carbon Energy. 2022;**4**:539-575

[132] Joshi B, Samuel E, Kim Y, il, Yarin AL, Swihart MT, Yoon SS. Progress and potential of electrospinning-derived substrate-free and binder-free lithium-ion battery electrodes. Chemical Engineering Journal. 2022;**430**:132876

[133] Jose Varghese R, Sakho EHM, Parani S, Thomas S, Oluwafemi OS, Wu J. Introduction to nanomaterials: Synthesis and applications. In: Nanomaterials for Solar Cell Applications. Amsterdam, Netherlands: Elsevier; 2019. pp. 75-95

[134] Nagata S, Atkinson GM, Pestov D, Tepper GC, Mcleskey JT. Electrospun polymer-fiber solar cell. Advances in Materials Science and Engineering. 2013;**2013**:975947

[135] Ponnamma D, Chamakh MM, Alahzm AM, Salim N, Hameed N, Al M. et al, Core-Shell nanofibers of Polyvinylidene fluoride-based nanocomposites as piezoelectric Nanogenerators. Polymers. 2020;**12**(10):2344

[136] Vu TH, Thu Nguyen H, Fastier-Wooller WJ, Tran CD, Nguyen TH, Nguyen HQ, et al. Enhanced Electrohydrodynamics for electrospinning a highly sensitive flexible fiber-based piezoelectric sensor. ACS Applied Electronic Materials. 2022;**4**(3):1301-1310

[137] Ding Y, Wang Y, Li B, Lei Y. Electrospun hemoglobin microbelts based biosensor for sensitive detection of hydrogen peroxide and nitrite. Biosensors & Bioelectronics. 2010;**25**(9):2009-2015

[138] Demirci Uzun S, Kayaci F, Uyar T, Timur S, Toppare L. Bioactive surface design based on functional composite electrospun nanofibers for biomolecule immobilization and biosensor applications. ACS Applied Materials & Interfaces. 2014;**6**(7):5235-5243

[139] Halicka K, Cabaj J. Electrospun nanofibers for sensing and biosensing applications—A review. International Journal of Molecular Sciences. 2021;**22**(12):6357

[140] Mercante LA, Pavinatto A, Pereira TS, Migliorini FL, dos Santos DM, Correa DS. Nanofibers interfaces for biosensing: Design and applications. Sensors and Actuators Reports. 2021;**3**:100048

[141] Deeraj BDS, Jayan JS, Saritha A, Joseph K. Electrospun biopolymer-based hybrid composites. In: Hybrid Natural Fiber Composites Material Formulations, Processing, Characterization, Properties and Engineering Applications. Sawston, Cambridge: Woodhead Publishing, Elsevier; 2021. pp. 225-252

[142] Li X, Li Z, Wang L, Ma G, Meng F, H. Pritchard R, et al. Low-voltage continuous electrospinning patterning. ACS Applied Materials & Interfaces. 2016;8(47):32120-32131.

[143] Bedford MN, Steckl JA. Photocatalytic self cleaning textile fibers by coaxial electrospinning. ACS Applied Materials & Interfaces. 2010;**2**(8):2448-2455

[144] Akdere M, Schneiders T. Modeling of the electrospinning process. In: Advances in Modeling and Simulation in Textile Engineering. Sawston, Cambridge: Woodhead Publishing, Elsevier; 2021. pp. 237-253

Chapter 2

# Selection and Fabrication of HMMC (AL6063-SIC-B4C-MG)

*Gurpreet Singh Matharou and Simran Kaur*

## Abstract

This unit deals with the selection and fabrication of HMMC (Al6063-10SiC-5$B_4$C-Mg) constituents by extensive biography review and satisfactory fabrication design. Researchers have promoted an extensive collection of Al6063 composites employing organic and inorganic reinforcements. The fundamental purpose of the broken-up stages is to constrain the metal matrix in a relevant capacity to strengthen the properties of the base materials. In the case of Al6063, the reinforcement weighty subject matter in the composite varies from 5 wt.% to 30 wt.%. Diverse classes of reinforcements had sought to integrate and operate in the composite formulation as hybrid reinforcements. This chapter further discusses the comprehensive development stages of 84% wt of Al 6063, 10% wt of SiC, 5% wt of $B_4C$ with 1% wt of Mg hybrid metal matrix composite (HMMC) through the stir casting approach. During the stir casting process, the melting action of the material emanates numerous gases and residuals apart from the expected composite. The residuals have numerous environmental concerns, which require discussion since some of the vapors and substantial waste can lead to detrimental effects on the environment in terms of air and soil pollution.

**Keywords:** hybrid metal matrix composite, stir casting, aluminum composite, environment

## 1. Introduction

**Figure 1** displays the flowchart of the HMMC development. During stir casting, a non-homogeneous mixture pattern has been an apprehension. The inclination is due to inappropriate segregation of reinforcement because of incorrect process parameters (rotation of stirrer, angle of stirring application, condition of wetting, and density). The material properties likewise have been reported to modify the characteristics of the homogeneous mix. The main metal matrix melted to obtain a molten state by melting it above its liquid temperature. The preheated reinforcement material is combined gradually so that a semi-solid-state is achieved. Repeatedly, the entire mix is needed to get heated to produce a molten state, and in between, stirring is done to attain the entirely conceivable consistency. The capability of the stir casting method predominantly rests on stirring speed, stirring duration, and stirring temperature [1, 2].

Here a crucible composed of ceramic or graphite is being utilized to melt the parent metal in a furnace. A mechanized stirrer with a graphite impeller with a rotation speed of around 150–800rpm is engaged to agitate the melt (see **Figure 2**) periodically. The

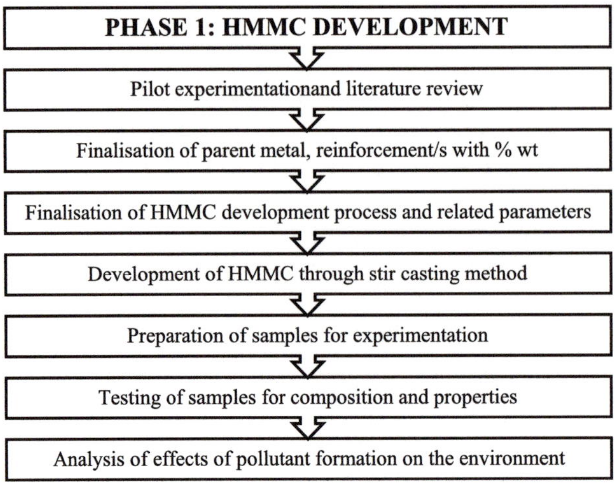


**Figure 1.**
*Flow-chart of HMMC development.*

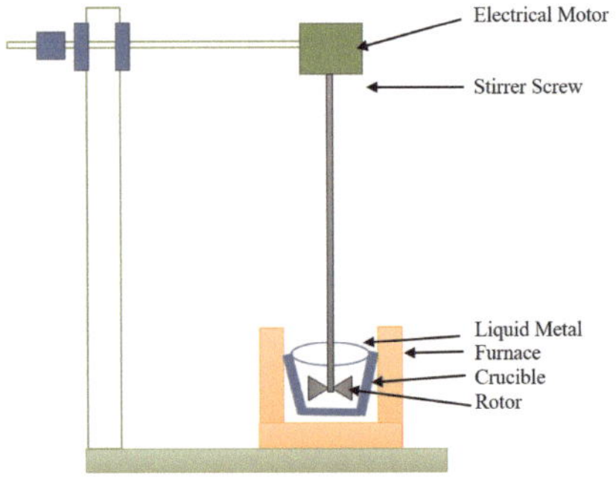


**Figure 2.**
*Stir casting technique for the MMCs fabrication [3].*

reinforcement materials are preheated to eliminate the humidity substances, facilitating wettability during stirring. Sakthivelu [4] had recommended a maximum limit of 30% of reinforcement for stable composites.

### 1.1 Stirring speed

The uniform dissemination of the reinforcement materials in the parent metal is essential for the advance in the coveted properties such as stiffness, toughness, tensile strength, etc. The stirrer with inadequate rpm provides an ineffectual activating force on the central metal matrix, contributing to an inadequate association [5]. Cluster arrangement and agglomeration inclination were recorded at slow rpm of mixing. The stirrer operated at high rpm provides considerable benefits in the creation of the expected composite since, at high rpm, the shear force supports the reinforced material to get the transfer inside the metal matrix dispersed phase and better bonding action with the metal matrix deep inside it, thereby setting up a coherent mix [5]. It has also been reported that porosity inclination can be stepped up at enhanced stirrer speed since gas particles induce inside the matrix.

### 1.2 Stirring duration

Stirring time likewise influences the distribution of dispersion into the metal matrix. Clustering of the material is observed at the lower stirring time, and further non-uniform mixture with fewer inclusions of reinforcement materials [6].

### 1.3 Stirring temperature

With a rise in temperature of the matrix metal, the viscosity was established to reduce, generating an effect in the reinforcement materials distribution. In extension, the chemical reaction was further revealed to develop with a rise in the temperature of the molten material [7].

## 2. Preparation of HMMC materials

The development of HMMC through stir casting typically uses the following phases. **Figure 3** illustrates the phases of melting of metal matrix composite to its melting point. The stirring of molten metal is managed to utilize an electric motor.

**Table 1** displays the stir casting process parameters retained during the fabrication of HMMC.

The reinforcement material is delivered with continuous stirring movement through a stirrer to associate the reinforcements in the matrix of the parent metal. The mixture is eventually poured into the mold and solidifies naturally. The pertinent equipment employed for HMMC development is summarized.

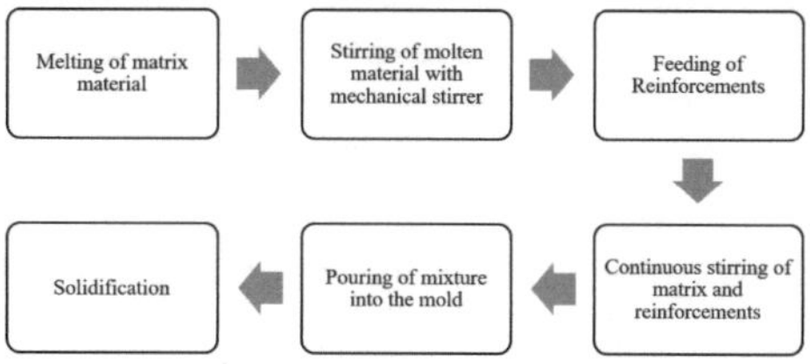


**Figure 3.**
*Phases of the stir casting process.*

| S. No. | Parameters | Values |
|---|---|---|
| 1 | Preheating chamber temperature | 850°C |
| 2 | Furnace temperature | 900°C |
| 3 | Core temperature | 750-800°C |
| 4 | Voltage | 440 V |
| 5 | Frequency | 50 Hz |
| 6 | Stirrer speed | 300–400 rpm |
| 7 | Die pre-heating temperature | 200°C |

**Table 1.**
*Stir casting process parameters.*

### 2.1 Furnace

**Figure 4** presents the original furnace adopted for the development of the HMMC. It has a temperature gauge with a regulator switch to regulate the temperature. The maximum temperature obtainable is around 1400°C. A convenient mechanical stir system generates a vortex in the melt, facilitating an exquisite melt blending, composing the metal matrix and related reinforcements. In order to evade the chances of solid particles settlement at the base of the crucible, a bottom pouring furnace is likewise suggested.

### 2.2 Mechanical stirrer rotor

The mechanical stirrer plays an essential role in forming an acceptable vortex in the melt to bring about the best possible coherence. Distinctive impeller stirrers can be used, i.e., single, double, and multiblade impeller. The double blade impeller (**Figure 5**) is employed mainly to develop AMCs. The single and multiblade impeller is handled primarily in chemical industries.

**Figure 6** displays impeller stirrers accepted for HMMC development. The blade was applied with a coating of zirconia onto a stainless-steel stirrer. The zirconia layer helps in averting probable reactions between the molted aluminum material and stainless steel of the stirrer. The impellers have been investigated for developing a sufficient vortex during the mixing process.

### 2.3 Crucible

Crucible is a container in which the metal matrix is melted to its molten temperature and the desired refracting materials are being added. Nowadays, diverse materials consisting of Alumina, Tungsten, Graphite, etc. are being adopted as a crucible. For HMMC development, the reinforcement materials (SiC and $B_4C$) were pre-heated in the Alumina crucible (**Figure 7**), whereas the parent metal (Al6063) is melted in a graphite crucible shown in **Figure 8**.

The Graphite crucible experiences the following advantages.

**Figure 4.**
*Electric furnace used for HMMC development.*

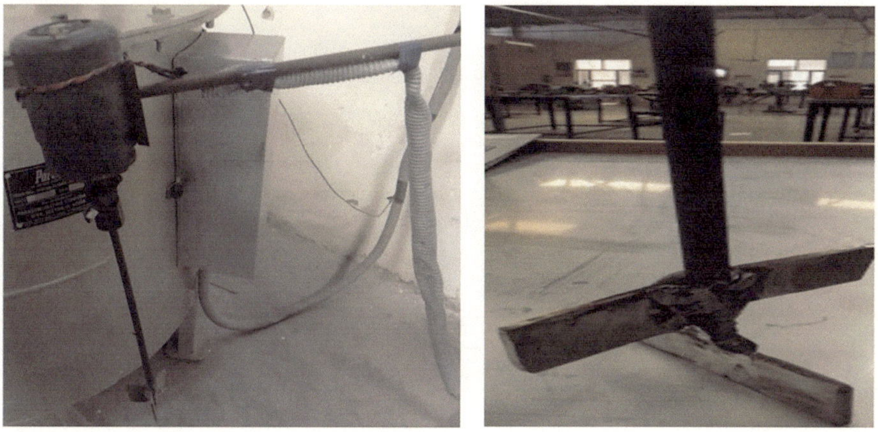

**Figure 5.**
*Actual photo of mechanical stirrer used.*

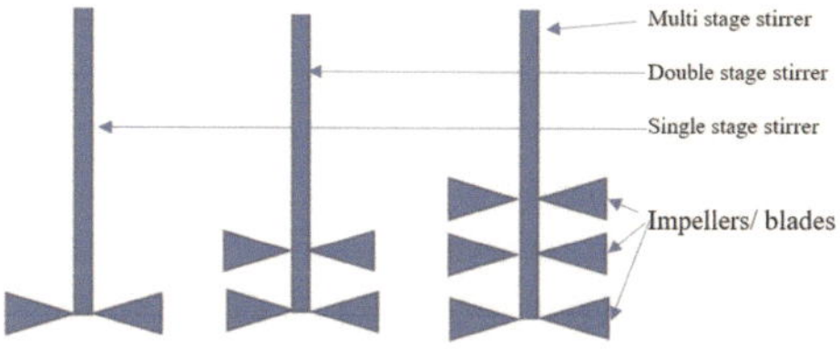


**Figure 6.**
*Impeller stirrer and types.*

**Figure 7.**
*Alumina crucible.*

1. High melting temperature (2500°C).

2. It is easily accessible.

3. The cost is less in comparison to tungsten.

4. Graphite has good electrical conductivity.

**Figure 8.**
*Graphite crucible.*

## 2.4 Power supply

The induction resistance furnace with a temperature regulator is linked with a three-phase electricity supply. To control the current and voltage supply, an ammeter and voltmeter were associated with the circuit. **Figure 9** illustrates the ammeter and voltmeter. The ammeter indicates the instantaneous current flowing in the circuit. The induction resistance furnace is engaged with moving iron type (M-Tech industries) with range 0-10A. The Voltmeter is likewise utilized for measuring the instantaneous voltage value across the circuit with a range 0–300 volt.

The current drawn by the electrical inductor furnace, depends on the furnace size, shape, and capabilities. The furnace shown withdraws around 55A to 75A. The efficiency spectrum of an electrical furnace is surprisingly modest; all modern electrical furnaces have an AFUE of 100%. That means that entire electric furnaces convert electricity into heat energy without any losses. Due to energy losses in ducts and the energy required to run a blower, the electric furnace is slightly expensive for operation [8]. The energy requirement is AC 380/7kw/50 Hz. The induction furnace also comprises a temperature regulator and digital display unit of temperature. **Figure 10** shows the Digital display unit with a regulator switch.

**Figure 9.**
*Ammeter and voltmeter.*

**Figure 10.**
*Digital display unit with regulator switch.*

### 2.5 Die

A graphite material die was utilized to shape and solidify molten material obtained after the rigorous stirring of Al 6063, SiC, and $B_4C$. The size of the die is $100 \times 50 \times 30$ with a tapered shape. **Figure 11** displays the die adopted for fabrication.

## 3. Steps accepted for HMMC fabrication

**STEP 1:** The stir casting setup employed for fabricating 84% Al-10% SiC-5% $B_4C$-1% Mg is shown in **Figure 12**. It consists of a furnace with a temperature range up to 14,000°C for heating the materials utilizing electrical resistance heating, which is the generally used technique of heat development. The mechanical stirrer fixed to the motor was supported inside the graphite crucible.

**STEP 2:** A correct weight measurement of the constituents is determined utilizing an electronic balancer (see **Figure 13a-c**). The silicon carbide and boron carbide

**Figure 11.**
*Graphite die with HMMC brick.*

**Figure 12.**
*Electric furnace.*

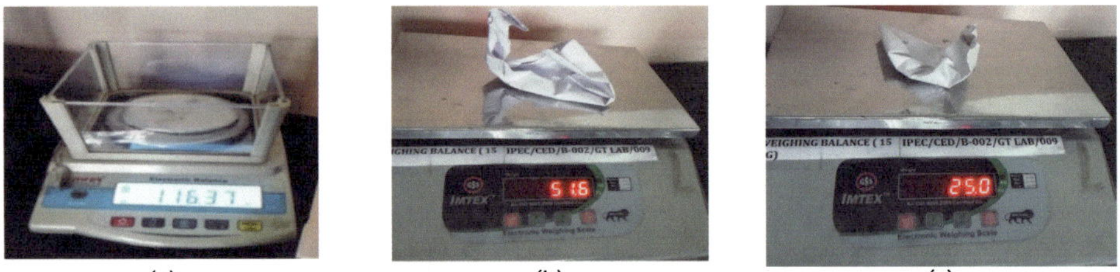

(a) (b) (c)

**Figure 13.**
*Wt. measurement of (a) Al 6063 material, (b) SiC (c) $B_4C$.*

particles were preheated at 850-900°C to eliminate any traces of moisture and oxidize their surface, forming a silicon oxide (SiO2) layer (see **Figure 14**). This layer enhances the wettability of the composite [9].

**STEP 3:** The Al 6063 billets were later melted in a new graphite crucible (see **Figure 15**). The temperature of the furnace was controlled between 850 and 950°C. A less slag was observed on the edges and was cut out by a graphite spoon. A uniform temperature of 700-750°C was secured, and flux was included in the melt to restrain oxidation [10].

**STEP 4:** The molten metal was later cooled naturally to a semi-solid state at around 550-600°C, and gradually preheated SiC, and $B_4C$ was included in the melt in fragments, and mechanical stirring was performed at 300–400 rpm (**Figure 16**).

**STEP 5:** Less than 1% wt of magnesium (Mg) was also added to develop the wettability of the mixture [7]. The mix was continuously stirred for 5 minutes, and the consistent mixture was poured into the die. **Figure 17** exhibits the graphite die along with the developed HMMC brick.

**STEP 6**: The solidified HMMC brick (**Figure 18**) is taken out from the die and is processed and cut into suitable identical pieces (30 × 20× 5 mm) further for experimentation. The specimens were then packed in polythene pouches with unique identification marks. Few specimens were then forwarded to authorized labs for SEM analysis and mechanical testing.

**Figure 14.**
*Preheating of SiC and $B_4C$.*

**Figure 15.**
*Aluminum billets melting in furnace.*

**Figure 16.**
*Motorized stirring of HMMC.*

**Figure 17.**
*Graphite die with HMMC brick.*

**Figure 18.**
*Solidified HMMC bricks.*

## 4. Samples of HMMC

**Figure 19(a)** and **(b)** show the weight measurement of brick one and brick two. The brick was cut into a smaller size of 30 mm × 20mm × 5 mm for experimentation on Die Sinking EDM. The wire EDM (Annexure 4) was used for preciously cutting the bricks so that the internal grain structures were not disturbed. **Figure 20** shows the line diagram of the HMMC sample, and **Figure 21** shows the actual sample.

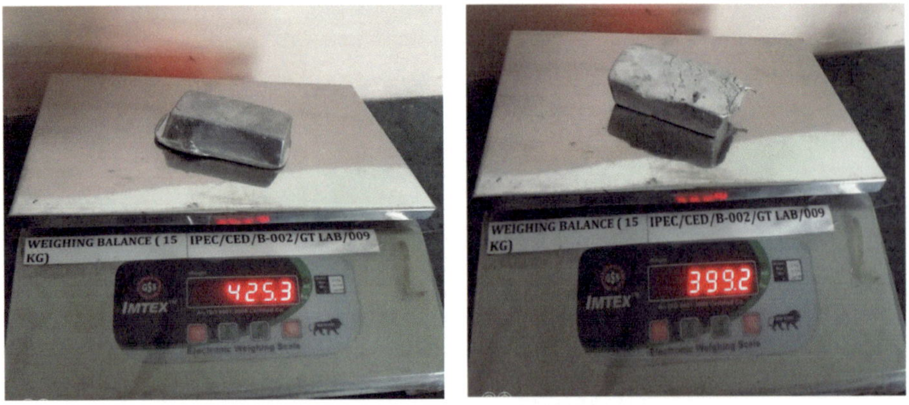


(a) (b)

**Figure 19.**
*(a). Wt. measurement of brick 1. (b). Wt. measurement of brick 2.*

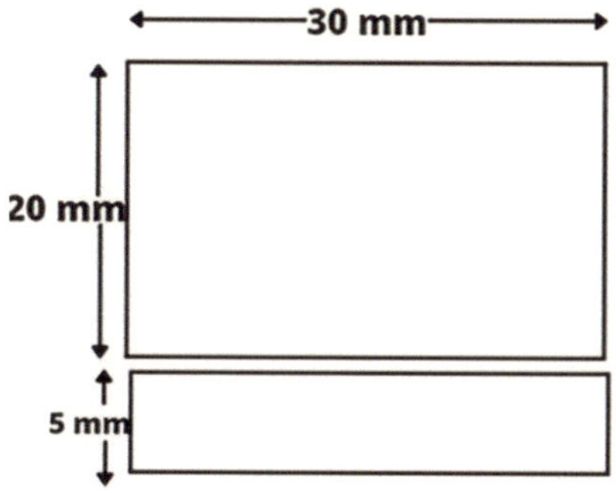


**Figure 20.**
*Line diagram of the specimen.*

**Figure 21.**
*View of the sample.*

# 5. HMMC properties and test analysis

## 5.1 Properties of the individual constituents

Aluminum 6063 is broadly employed as a general-purpose alloy in many engineering applications such as the extrusion process, owing to its fair strength [11]. **Table 2** exhibits the constituents of Al6063.

**Table 3** (a-c) presents the physical and thermal properties of $B_4C$ [12], Al6063 [2], and SiC [13], respectively.

Boron carbide ($B_4C$) is one of the hardest materials available. Above 1250°C, it has been harder than cubic boron nitride and diamond. $B_4C$ is an alluring reinforcement substance owing to its unique balance between thermal and chemical properties. Moreover, it possesses a smaller density and greater hardness

| Al6063 | Wt. % | Al6063 | Wt. % |
|---|---|---|---|
| Al | Max 97.5 | Mn | Max 0.2–0.7 |
| Cr | Max 0.11 | Si | 0.2–0.9 |
| Cu | Max 0.09 | Ti | Max 0.11 |
| Fe | Max 0.36 | Zn | Max 0.92 |
| Mg | 0.44–0.89 | | |

**Table 2.**
*Composition of Al6063 [2].*

| (a) $B_4C$ | | (b) SiC | | (c) Al 6063 | |
|---|---|---|---|---|---|
| Properties | Value | Properties | Value | Properties | Value |
| Specific Heat (°C) | 700 | Coefficient of Thermal Expansion (°C) | 4 | Coefficient of Thermal Expansion (per ° C) | 0.000022 |
| Melting Point (°C) | 2783 | Specific Heat (°C) | 750 | Thermal Conductivity (cal/cm$^2$/ cm/ Celsius at 25°C) | 0.285 |
| Density (g/cm$^{3)}$ | 2.55 | Melting Point (°C) | 2730 | Electrical Conductivity (% copper standard at 20 °C) | 33.5 |
| Thermal Conductivity (W/mk) | 17–42 | Density (g/cm$^{3)}$ | 3.21 | Density (g/cm$^3$) | 2.64 |
| Hardness (Kg/mm$^2$) | 2900–3580 | Thermal Conductivity (W/mk) | 120 | Freezing Range (°C) approx. | 625–525 |

**Table 3.**
*(a, b) physical properties of $B_4C$ and SiC (c) thermal properties of Al6063.*

value of order 30 GPa. Thus, $B_4C$-reinforced HMMCs fabricated through the moderate-cost stir casting structure have gained higher attractiveness among researchers [14, 15]. $B_4C$ has good mechanical strength with desired properties of neuron absorption [16].

Silicon carbide (SiC) is constituted of tetrahedra of silicon and carbon atoms with influential bonds in the crystal lattice. The SiC material has less thermal enlargement, immense strength, and thermal conductivity of greater order and has been recorded to be resistant against thermal shock [17, 18]. The SiC can tolerate severe temperatures and has got high hardness coupled with low density.

Magnesium (Mg) is acknowledged for promoting grain refinement, wettability, and reinforcing the solid solution [19].

## 5.2 Properties of the HMMC

The spark atomic emission spectrometry (SAES) was conducted with ASTM E1251–11 standards (test procedure for Al and Al alloys) to determine the elements present in the HMMC samples. **Table 4**(a) illustrates the composition and %wt of elements. **Table 4** (b) indicates the HMMC significant properties. The density of HMMC (2637 kg/m$^3$) as obtained through the test report has been used to calculate the MRR and EWR [20, 21].

## 5.3 SEM analysis of the HMMC

For establishing the homogeneity of the HMMC, the sample was tested by employing scanning electron microscopy (SEM). **Figure 22** displays the uniform dispersion of SiC and $B_4C$ in the specimen. No segregation of SiC grains along with $B_4C$ particles was stationed along the grain edges. Dissemination of grains is acknowledged to be intra-granular, in which the maximum particles locate inside the grains. The uniform distribution is commensurate with the efficient and timely stirring action during the stir casting process [22]. The crater's size is less with $B_4C$ particles this could be because of the creation of a boron oxide ($B_2O_3$) layer on the $B_4C$ ceramic, because of

| (a) | | | | (b) | |
|---|---|---|---|---|---|
| **HMMC** | **% Wt** | **HMMC** | **% Wt** | **Properties** | **Value** |
| Al | 91.43 | Mg | 0.087 | | |
| Si | 4.151 | Cr | 0.037 | Melting point | 750–800 |
| Cu | 1.516 | Ti | 0.027 | U.T.S | 110Mps |
| Zn | 1.498 | Sn | 0.022 | Tensile Strength | 118 Mpa |
| Fe | 0.806 | V | 0.006 | Break Load | 9.44KN |
| Mn | 0.149 | Cd | 0.0019 | Yield Stress | 82 Mps |
| Ni | 0.13 | Co | <0.001 | Hardness | 72HB |
| Pb | 0.089 | | | Density | 2637 kg/m$^3$ |

**Table 4.**
*(a) Composition of HMMC (b) significant properties of HMMC.*

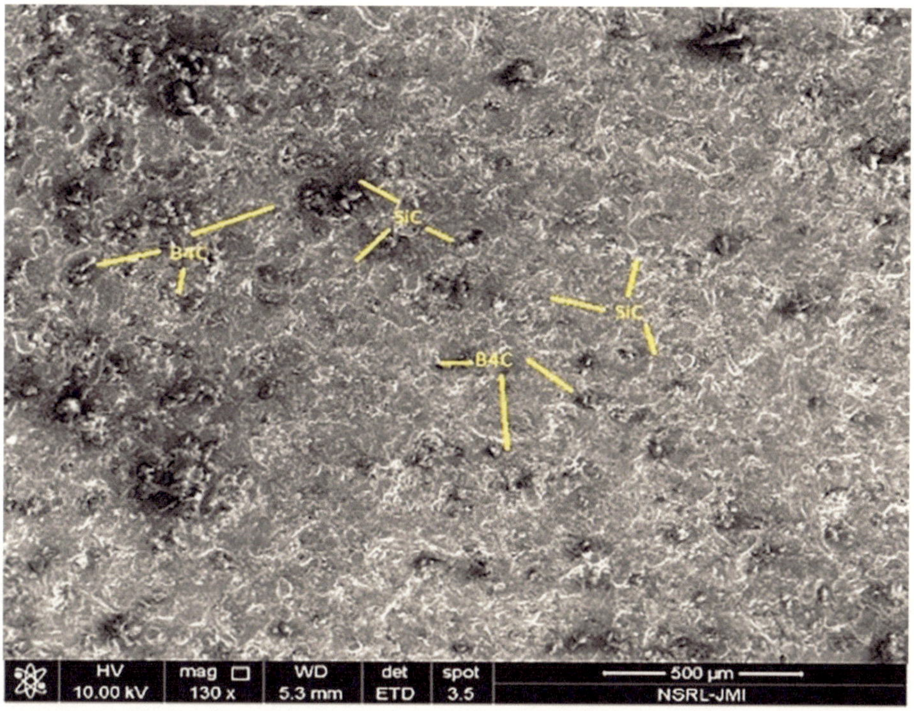


**Figure 22.**
*SEM image of the stir casted HMMC.*

liquid-to-liquid reaction leading to an expansion in the wettability, which is observable at a specialized high temperature [23]. Many researchers proposed that reinforcement in the particulate form up to wt. 30% may be included in a molten metal matrix to perform a more substantial reinforcement distribution [24]. Reinforcement is added emphatically into the molten stage of aluminum. The stirring speed, time of stir, stirrer blade angle, pouring temperature, solidification rate, reinforcement size, and elements percentage influence the fabricated composite consistency.

| Sr. No | Unwanted Gases/ Residual Waste | Environmental concerns and human health |
|---|---|---|
| 1 | Aluminum hydroxide | Exposure to Aluminum hydroxide may cause repulsion, vomiting, hyperacidity, pungency, Low blood phosphates (hypophosphatemia), distaste, causticity leading to bowel obstruction, Fecal impaction [25]. |
| 2 | Aluminum oxides | Indicative toxicity has been reported, followed by chronic inhalation of the aluminum oxides. Long-term aluminum oxide inhalation may cause pneumoconiosis with cold and exertion and a restrictive pattern of rib cage function. In severe cases, death has been reported due to respiratory failure. |
| 3 | Aluminum sulfates | Eating or gulping aluminum sulfate produces serious disturbance to the digestive organs and stomach. An influenced individual may encounter retching, queasiness, and runs, adding water to aluminum sulfate can make a sulfuric acid structure. The sulfuric acid may cause soil damage by reducing its constituents. |
| 4 | Boron Oxides | Acute effects: The boron oxides contacts can aggravate the skin and eyes. Breathing in Boron Oxide can bother the nose and throat, causing hacking and wheezing. Introduction to Boron Oxide may cause heaving wooziness, cerebral pain, sickness, and so forth. |
| | | Chronic Effects: The accompanying long-haul wellbeing influences may happen after some time getting an introduction to Boron Oxide and can keep going for months to years [26]. Boron oxide may make permanent damage to kidney and livers. |
| 5 | Silicon Dioxide | Silicon Dioxide exists naturally on earth and our bodies. No evidence has been reported to advocate its implication on human health, but more research is required to ascertain its role on the body. Inhalation of silica dust may cause diseases related to breathing. |
| 6 | Fly Ash particles | It can get placed in the deepest part of the lungs, where it may cause an asthmatic attack, inflammation, and immunological reactions. They contribute to Particulate Matter 2.5 and 10. |
| 7 | Magnesium Oxide | Exposure to Magnesium Oxide can cause "metal fume fever" which is a symptom in which the patent gets a metallic taste in the throat with headache, sneezing symptoms, cold symptoms. |

**Table 5.**
*List of unwanted gases/residual waste/effects on the environment.*

## 6. Environmental concerns

The stir casting process involves melting the metal at around 800-1000°C. The metal matrix used is Aluminum 6063 with Boron Carbide ($B_4C$) and Silicon Carbide (SiC) as reinforced materials. The melting operation produces specific unwanted gases and residual waste, which must be discussed. **Table 5** illustrates the relevant unwanted gases/residual waste with apprehensions on the environment and human beings [27].

## 7. Summary

This chapter focuses on the comprehensive development stages of the HMMC (84%wt of Al 6063–10%wt of SiC-5%wt of $B_4C$ with 1%wt. of Mg) through the stir casting method. A comprehensive description of the essential ingredients (electric furnace, stirrer, the crucible, die) required for HMMC fabrication and the procedures has been covered. Reinforcement is added emphatically into the molten stage of aluminum. The stirring speed, time of stir, stirrer blade angle, pouring temperature, solidification rate, reinforcement size, and elements percentage influence the fabricated composite consistency. The developed HMMC was further analyzed for composition and specific mechanical and thermal tests. The HMMC density of 2700 $kg/m^3$ was noted for MRR calculations. For confirming the homogeneity of the HMMC, the sample was analyzed using an SEM test. Dissemination of grains was noticed to be intra-granular, in which the maximum particles reside inside the grains. The uniform distribution is proportional due to the efficient and timely stirring action during the stir casting process. The crater's size is observed to be less with $B_4C$ particles because of the creation of a boron oxide ($B_2O_3$) layer on the $B_4C$ ceramic because of liquid-to-liquid reaction leading to an expansion in the wettability, which is observable at a specialized high temperature. During the stir casting process, the melting action of material emits out certain gases and residuals apart from the required composite. The residuals have specific environmental concerns. The severe effects caused by aluminum hydroxide, aluminum oxide, aluminum sulfate, boron oxide, silicon dioxide, magnesium oxide, and fly ash on the environment have also been covered.

## Author details

Gurpreet Singh Matharou and Simran Kaur*
Manav Rachna International Institute of Research and Studies, Faridabad, Haryana, India

*Address all correspondence to: simran.foc@mriu.edu.in

## References

[1] Kareem A, Qudeiri JA, Abdudeen A, Ahammed T, Ziout A. A review on AA 6061 metal matrix composites produced by stir casting. Materials. 2021;**14**:175. DOI: 10.3390/ma14010175

[2] Matharou GS, Bhuyan BK. Experimental investigation of surface roughness in electric discharge machining of hybrid metal matrix composite. In: Govindan K, Kumar H, Yadav S, editors. Advances in Mechanical and Materials Technology. Vol. 20. Singapore: Springer; 2022. pp. 333-343. DOI: 10.1007/978-981-16-2794-1_30

[3] Mishra D, Tulasi T. Experimental Investigation on Stir Casting Processing and Properties of Al 6082/SiC Metal Matrix Composites. In: GSVL N, Babu AV, Reddy SS, Dhanasekaran R, editors. Recent Trends in Mechanical Engineering. Vol. 14. Singapore: Springer; 2020. pp. 159-168. DOI: 10.1007/978-981-15-1124-0_14

[4] Sakthivelu S, Sethusundaram PP, Meignanamoorthy M, Ravichandran M. Synthesis of metal matrix composites through stir casting process – a review. Mechanics and Mechanical Engineering. 2018;**22**:357-370. DOI: 10.2478/mme-2018-0029

[5] Mehta VR, Sutaria MP. Investigation on the effect of stirring process parameters on the dispersion of SiC particles inside melting crucible. Metals and Materials International. 2020;**27**: 2989-3002. DOI: 10.1007/s12540-020-00612-0

[6] Sahu MK, Sahu RK. Fabrication of aluminum matrix composites by stir casting technique and stirring process parameters optimization. In: Vijayaram TR, editor. Advanced Casting Technologies. London, UK: InTech; 2018. DOI: 10.5772/intechopen.73485

[7] Sozhamannan GG, Prabu SB, Venkatagalapathy VSK. Effect of processing Paramters on metal matrix composites: Stir casting process. JSEMAT. 2012;**02**:11-15. DOI: 10.4236/jsemat.2012.21002

[8] Ravichandran M, Meignanamoorthy M, Chellasivam GP, Vairamuthu J, Kumar AS, Stalin B. Effect of stir casting parameters on properties of cast metal matrix composite. Materials Today: Proceedings. 2020;**22**:2606-2613. DOI: 10.1016/j.matpr.2020.03.391

[9] Bodukuri AK, Eswaraiah K, Pradeep V. Investigation on machining of hybrid metal matrix composite. MSF. 2019;**969**: 846-851. DOI: 10.4028/www.scientific.net/MSF.969.846

[10] Bains PS, Sidhu SS, Payal HS. Fabrication and machining of metal matrix composites: A review. Materials and Manufacturing Processes. 2016;**31**: 553-573. DOI: 10.1080/10426914.2015.1025976

[11] Venkatesulu M, Rama Kotaiah K. Production and mechanical properties of AL 6063/B4C composites. JMERD. 2019; **42**:46-49. DOI: 10.26480/jmerd.01.2019.46.49

[12] Gudipudi S, Nagamuthu S, Subbian KS, Chilakalapalli SPR. Enhanced mechanical properties of AA6061-B4C composites developed by a novel ultra-sonic assisted stir casting. Engineering Science and Technology, an International Journal. 2020;**23**:1233-1243. DOI: 10.1016/j.jestch.2020.01.010

[13] Das S, Acharya U, Rao SVVNS, Paul S, Roy BS. Assessment of the surface

characteristics of aerospace grade AA6092/17.5 SiCp-T6 composite processed through EDM. CIRP Journal of Manufacturing Science and Technology. 2021;**33**:123-132. DOI: 10.1016/j.cirpj.2021.03.005

[14] Patidar D, Rana RS. Effect of B 4 C particle reinforcement on the various properties of aluminium matrix composites: A survey paper. Materials Today: Proceedings. 2017;**4**:2981-2988. DOI: 10.1016/j.matpr.2017.02.180

[15] Toptan F, Kilicarslan A, Karaaslan A, Cigdem M, Kerti I. Processing and microstructural characterisation of AA 1070 and AA 6063 matrix B4Cp reinforced composites. Materials & Design. 2010;**31**:S87-S91. DOI: 10.1016/j.matdes.2009.11.064

[16] Naidu VVB, Varaprasad KC, Prahlada Rao K. Machinability analysis on wire electrical discharge machining of stir casted AA2024/Al 2 O 3/BN hybrid composite for aerospace applications. Materials and Manufacturing Processes. 2021;**36**:730-743. DOI: 10.1080/10426914.2020.1854466

[17] Pul M. Effect of sintering on mechanical property of SiC/B 4 C reinforced aluminum. Materials Research Express. 2018;**6**:016541. DOI: 10.1088/2053-1591/aacee1

[18] Sivananthan, S., Ravi, K., Samson Jerold Samuel, C., 2020. Effect of SiC particles reinforcement on mechanical properties of aluminium 6061 alloy processed using stir casting route. Materials Today: Proceedings 21, 968–970. doi:10.1016/j.matpr.2019.09.068

[19] Suneesh E, Sivapragash M. Comprehensive studies on processing and characterization of hybrid magnesium composites. Materials and Manufacturing Processes. 2018;**33**:1324-1345. DOI: 10.1080/10426914.2018.1453155

[20] Matharou GS, Bhuyan BK. Parametric optimization of EDM processes for aluminum hybrid metal matrix composite using GRA-PCA approach. International Journal of Mechanical and Production Engineering Research and Development. 2020; 10: 367–378. DOI: 10.24247/ijmperdjun202034.

[21] Matharou GS, Bhuyan BK. Modelling and combined effect analysis of electric discharge Machining process using response surface methodology. Materials Today: Proceedings. 2021;**46**:6638-6643. DOI: 10.1016/j.matpr.2021.04.103

[22] Senthilkumar TS, Muralikannan R. Role of TiC and h-BN particles on morphological characterization and surface effects of Al 4032 hybrid composites using EDM process. Journal of Mechanical Science and Technology. 2019;**33**:4255-4264. DOI: 10.1007/s12206-019-0822-z

[23] Bystrenko O, Jiang J, Dong F, Li X, Qiu J, Liu J, et al. Kinetics of bonds at structural breakdown in boron carbide under intensive loads: A molecular dynamics study. Computational Materials Science. 2020;**180**:109711. DOI: 10.1016/j.commatsci.2020.109711

[24] Garg P, Jamwal A, Kumar D, Sadasivuni KK, Hussain CM, Gupta P. Advance research progresses in aluminium matrix composites: Manufacturing & applications. Journal of Materials Research and Technology. 2019;**8**:4924-4939. DOI: 10.1016/j.jmrt.2019.06.028

[25] Bahrami A, Soltani N, Pech-Canul MI, Gutiérrez CA. Development of metal-matrix composites from industrial/agricultural waste materials

and their derivatives. Critical Reviews in Environmental Science and Technology. 2016;**46**:143-208. DOI: 10.1080/10643389.2015.1077067

[26] Wang QG, Crepeau PN, Davidson CJ, Griffiths JR. Oxide films, pores and the fatigue lives of cast aluminum alloys. Metallurgical and Materials Transactions B. 2006;**37**:887-895. DOI: 10.1007/BF02735010

[27] Matharou GS, Bhuyan BK. Hybrid Metal Matrix Composite Development by Stir Casting and Environmental Concerns. Vol. 17. Singapore: Advances in Engineering Materials Springer; 2021. pp. 377-386. DOI: 10.1007%2F978-981-33-6029-7_35

Chapter 3

# Experimental Investigation of the Mechanical and Thermal Properties of Natural Green Fibres

*Dheeraj Kumar, Nadeem Faisal, Ramit Choudhury and Swarup S. Deshmukh*


## Abstract

Biomaterials and green products rely heavily on natural lignocellulosic fibres. They have a wide variety of potential capabilities and characteristics, making them suitable for many applications. These fibres offer all the components required for renewable energy deployment. Fibre polymers from Jharkhand such as palm, datura, lemon, and mustard were studied for their thermal, mechanical, and interfacial adhesion properties. There were also tests on tensile strength, elongation to break, and thermogravimetric analyses (TGA). The effects of heating on weight loss, water loss, and disintegration have also been studied. A comparison was made between frequently used global fibres and the fibres analysed in this research article. Jharkhand's fibres are shown to be more compromising than worldwide fibres. Palm fibres have excellent tensile strength (160 MPa) and modulus of elasticity (5 GPa). The thermal behaviour of lemon and datura fibres is the most similar. Palm and mustard fibres respond similarly in warm temperatures. At 140°C and 240°C, mass loss was 18.8 and 24.3%, respectively. TGA shows that the studied fibres are more suited for industrial applications owing to their stable thermal behaviour. Plastics, textiles, packaging, and papers may all use palm fibres in insulators, circuit boards, switches, and terminals, as well as in furniture and window frames.


**Keywords:** green materials, lignocellulosic fibres, mustard, eco-friendly, datura, palm, lemon

## 1. Introduction

New bio-product materials have been created to reduce dependency on petroleum commodities, potentially leading to sustainable green products and cleaner manufacturing [1]. Many countries have emphasised the use of bio-based renewable resources because of the rising cost of petroleum commodities, climate change, and the world's drive toward global sustainability [2]. Furthermore, in order to ensure long-term sustainability, the government has recognised the importance of available natural resources, their proper utilisation, and waste management, which has resulted in the development of better schemes, regulations, and promotions for natural bio-based materials, such as natural fibre composites (NFCs). Furniture, automotive, agriculture, construction, packaging, aerospace, and other industries have recently embraced natural fibre composites to replace conventional materials [3]. NFCs provide a number of advantages over traditional synthetic materials, including being recyclable, abundant, lighter in weight, degradable, and less expensive. NFCs also benefit from being a green product since they are recyclable and degradable in nature, contributing to the goal of environmental sustainability. Green commodities generated from agricultural waste would open the way for new renewable resources while also providing a source of revenue for many developing nations [4]. Plants (lignocellulosic) are natural fibres that may be used as polymer reinforcement, making them an excellent renewable resource. To develop new types of biomaterials that are ecologically friendly [5–7]. They might also be regarded a wonderful solution for reinforcement from an environmental and economic standpoint. They are more ecologically friendly, lighter, and require less energy, making them more sustainable from a sustainability aspect. Natural fibres are classified according to where they come from in the plant: leaf, fruit, seed fibres, stem fibres, bast fibres, and skin. Natural fibres are in great demand in a variety of sectors, including furniture, automotive, agricultural, construction, packaging, and aerospace, due to their high specific characteristics and cheap cost. Constituent qualities, maximum manufacturing temperature, degradability, orientation, volume fraction, fibre length, and geometry all have a role in the overall features and attributes of a bio-composites product. However, while designing green goods, mechanical and chemical qualities are the only two factors that are taken into account to a higher degree, since there is a correlation between natural fibre's chemical properties and their equal mechanical performance [8, 9].

Nowadays, lignocellulosic fibres are employed as a stand-alone and primary component of bio-composites. They are utilised separately because their inherent qualities are difficult to alter or modify in comparison to their equivalents, such as polymers, whose properties are readily manipulated. This might explain why natural fibres have less applicability in numerous industries than polymers, which have been widely employed in recent years. Natural fibres in bio-composites, on the other hand, might be useful in supplying alternative materials for green solutions [10].

This study presents a fresh approach to researching natural fibres accessible in India's Jharkhand area, particularly lignocellulosic fibres, and testing their mechanical and environmental behaviour from a variety of technical perspectives. This would ensure the creation of an ideal database of materials that could be crucial in developing green materials that are both eco-friendly and cost-effective, as well as developing future sustainable materials with broad industrial applications and opening doors for further bio-materials research.

## 2. Materials and methods

Diverse lignocellulosic wastes have been collected in abundance from several areas in Jharkhand (the north-eastern portion of India). In this work, we measured the physical properties of polymers.

Lignocellulosic fibre interface characteristics and thermogravimetric analyses were explored with the purpose of selecting lignocellulosic fibre.

### 2.1 Selection of fibres

The preliminary inquiry was carried out in a specified manner in accordance with assessment criteria in order to identify the best suited fibres. The basic selection for the compatibility of the different agricultural waste fibres was made. The availability of resources (fibres), cost estimate, dependable qualities necessary to complete the job (mechanical property), renewal duration, and material density are all included in this assessment criteria [11, 12]. A few species were omitted from further examination because they did not meet the original criterion. Lemons, palms, *Datura stramonium* (hence referred to as datura) and mustard residue fibres were used to determine the results. As previously stated, all of the fibres were readily accessible in large amounts in Jharkhand. As a result, they met the physical and mechanical requirements for the presentation while also being less expensive.

Following that, suitable fibre samples were investigated for mechanical and thermal examinations to determine their prospective capabilities. Samples were processed and adaptable in accordance with the test's requirements. The specimen to be tested for the tensile test, for example, was held at 120 mm in length with a gage length of 50 mm. Their diameter, on the other hand, has fluctuated based on the fibre variations and plant type. As a result, the averaged cross-sectional area estimates are based on the measurements recorded for each of the 10 possible gage length interpretations (for every 5 mm). Tensile strength, Young's modulus, and elongation obtained before the fracture point were all tested on all of the manufactured samples. Five testing trials were conducted for each fibre material, according to ASTM D 3822/01, and the average value was picked for future analysis. The 3365 Instron (**Figure 1**) is utilised for the tensile test, with a crosshead speed of 2.0 mm/min.

Furthermore, all of the fibres used in the study were from agricultural waste and were tested for thermal properties (**Table 1**). It was examined to determine the percentage loss of weight characteristics in touch with the combustion and the heating impact of the combustion. The conclusion is that they are environmentally beneficial materials for the manufacturing process.

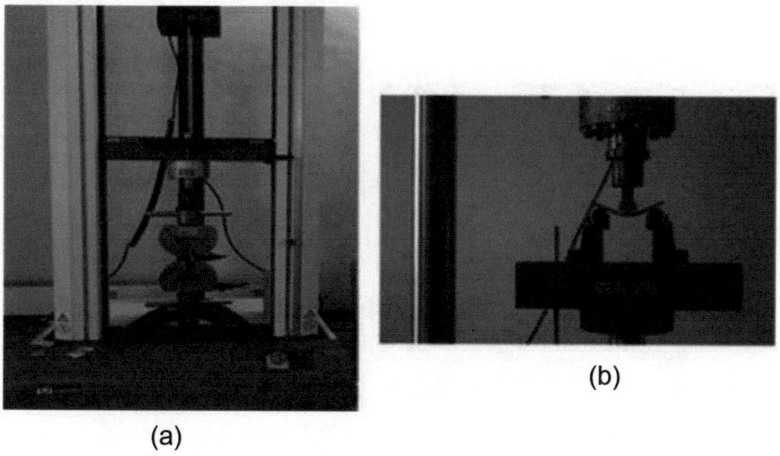

**Figure 1.**
*UTM Instron 3365. (a) Tensile, (b) flexural.*

| Polymers | Melt flow rate at 230°C and 2.2 kg load | Yield tensile strength (MPa) | Density ($kg/m^3$) |
|---|---|---|---|
| Polypropylene (PP) | 12.5 g/10 min | 34 | 906 |
| Low-density polyethylene (LDPE) | 0.85 dg/min | 11 | 921 |
| High-density polyethylene (HDPE) | 0.7 g/10 min | 30 | 960 |
| Polyvinyl chloride (PVC) | — | K value 68 | 569 |

**Table 1.**
*Physical properties of the selected polymers.*

## 2.2 Thermogravimetric analysis (TGA)

Thermogravimetric analysis (TGA) is used to identify changes in physical characteristics and biochemical processes that occur to fibres when the temperature is allowed to rise at the same rate as the temperature, since it is critical to retain information on fibres from agricultural wastes in terms of water loss and decomposition. TGA experiments are carried out with the aid of a NETZSCH TG 209/F1 apparatus (**Figure 2**). Thermal Stability was measured with several specimens at a heating rate of 10°C/min across the whole temperature range of 31–300°C and 500°C.

## 2.3 Characteristics of the interfacial zone

The pull-out technique is a useful tool for examining the interfacial characteristics of different polymer fibres. Because the properties of the fibre-matrix interface have a substantial influence on composite materials' mechanical performance. The interfacial effects of the debonding and pull-out processes may be determined using this approach. The fibre was put into the polymer and a tensile test was conducted in the pull-out inquiry. In this testing method, the polymer and the free tip of the fibre are

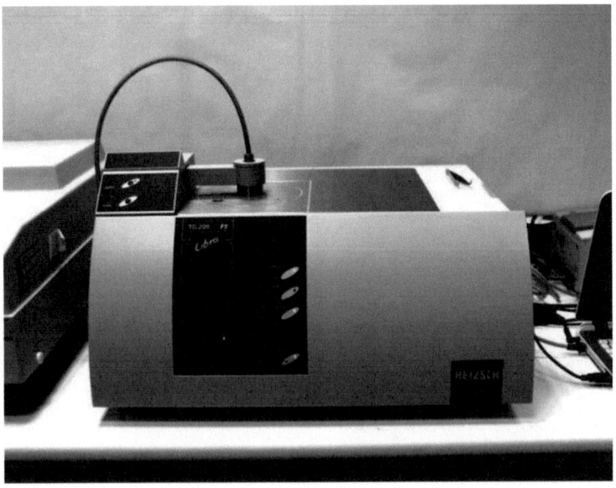

**Figure 2.**
*NETZSCH TG 209/F1.*

both gripped and pulled apart. The fibre was removed without causing the polymer to split. This data suggests that the fibre/polymer interface was a failure of adhesive rather than polymer or fibre cohesiveness. The durability and capacity of polymers or fibres were explored in order to build environmentally acceptable materials. For this aim, a pull-out technique is employed to determine the maximum load capacity that can sustain the fibre/polymer [13].

The extrusion procedure was chosen for the preparation of the pull-out sample's method of hot mixing. For each specimen, 31 cm of rectangles were cut. The sample is constructed such that each fibre is implanted along the rectangle's centre axis. An optical microscope is used to measure the specimen's length and diameter. With the aid of a universal testing equipment, the specimen is permitted to act a tensile load (Instron 3365). The load-displacement graph is being recorded, and the crosshead speed is being held at 2.0 mm/min.

## 3. Result and discussion

Various lignocellulosic fibres have been discovered in the Jharkhand region in this investigation. Environmental waste is a consequence of using green bio-based goods instead of conventional resources. Unfortunately, Jharkhand is not doing enough to make use of its agricultural lignocellulosic waste. The qualities of this fibre and the potential alternative uses of global fibres remain unknown to industrialists. In the creation of biomaterials for green goods, as well as the search for materials that will help in the finding of renewable sources of materials for green products, these fibres will be of use. As a result, developing countries and industrial markets like India and its neighbours will see an increase in their gross domestic product.

### 3.1 Mechanical investigation

A mechanical experiment was conducted to determine the mechanical characteristics of Jharkhand's agriculture waste fibres. The tensile testing procedure is used to determine Young's modulus, length elongation before fracture, and maximum tensile strength. To determine the tensile strength, all of the samples provided for the research have been finished. The potential possibilities of biomaterials have been discovered as a result of this research [14]. It is comparable to traditional fibres such as coir, sisal, flax, hemp, jute, and others that are often used in literature. **Figure 3** depicts the mechanical behaviour (stress-strain) of lignocellulosic fibres. The tensile strength of natural fibres is seen in **Figure 4**. Because of the larger cellulose concentration in its constituents, palm fibre has the most substantial tensile strength of 160 MPa, whereas mustard fibre has 60 MPa. Datura has the lowest value, which is less than 8 MPa. As a result, the tensile strength of fibres is arranged in the following order: palm, mustard, and lemon fibres.

**Figure 5** also depicts the elongation variation required to break the property of Jharkhand agro-based fibres. It demonstrates that palm fibres have the highest elongation value to break the percentage. Palm fibre, mustard, and datura, respectively, have 0.078%, 0.06%, and 0.026%. The lemon type fibre, on the other hand, has the lowest percentage value, at 0.025%. Except for the palm fibres, the previous observations of elongation to break the % value of mechanical qualities are extremely similar. If just a single criterion is used for the selection process, it may result in paying little attention to the other fibre materials if one characteristic is overlooked. As a result, it is

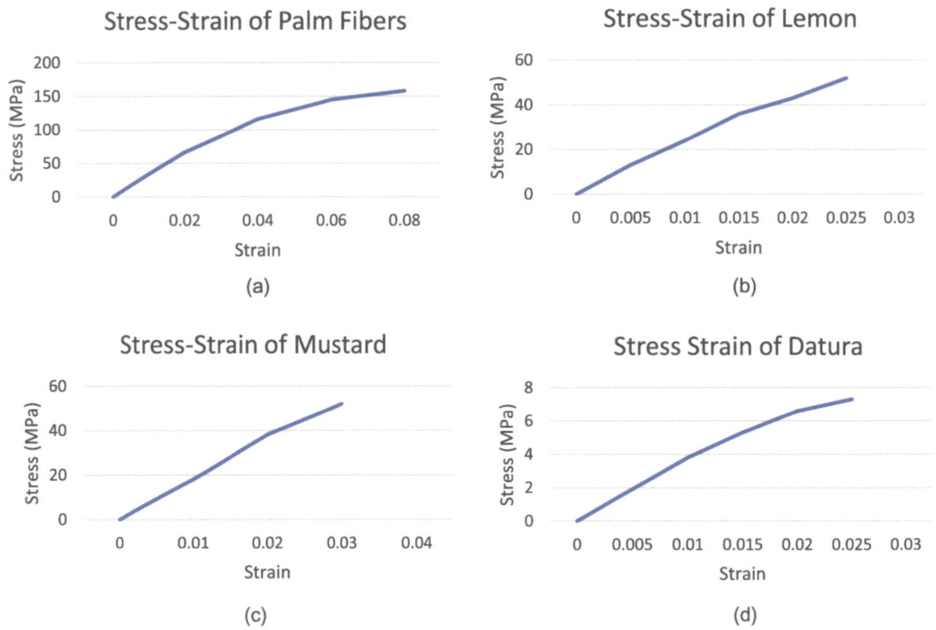


**Figure 3.**
*Stress-strain diagram of (a) palm fibres, (b) lemon, (c) mustard and (d) datura.*

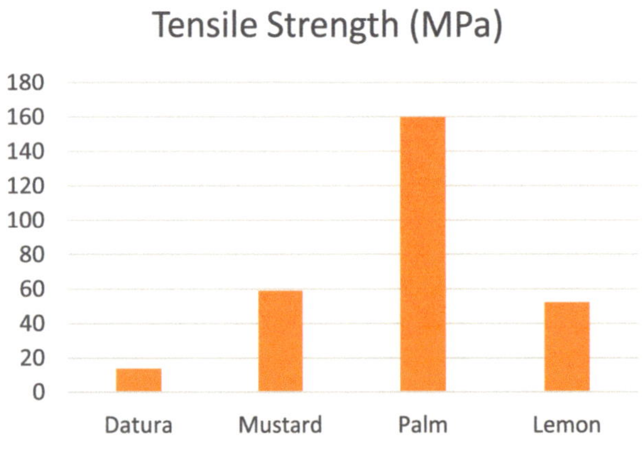


**Figure 4.**
*Tensile strength of fibres compared.*

concluded that more than one criterion should be used in the selection of natural fibre constituents for biomaterial evaluation.

Young's modulus criteria with mechanical parameters for Jharkhand agricultural waste fibres is shown in **Figure 6**. Palm fibres have a Young's modulus value of 5.02 GPa, whereas lemon fibres have a value of 2.71 GPa. Similarly, mustard-type fibres have a modulus of elasticity closer to 2.35 GPa, but datura only possesses 0.34 GPa. It was discovered that fibres from Jharkhand had a higher Young's modulus, indicating that they would make superior natural reinforced polymer composites.

The intrinsic capabilities of regularly used fibre materials and Jharkhand lignocellulosic fibres have been found. Young's modulus and tensile strength of Jharkhand

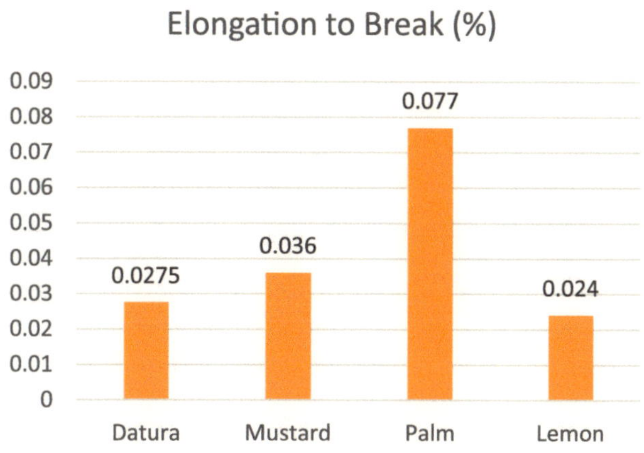


**Figure 5.**
*Elongation to break % of fibres compared.*

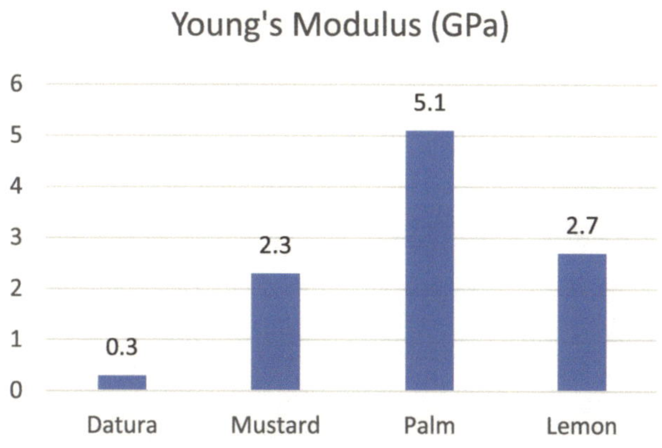


**Figure 6.**
*Young's modulus % of fibres compared.*

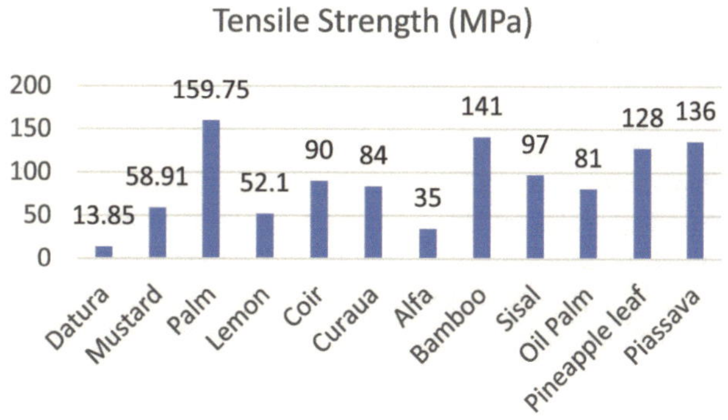


**Figure 7.**
*Tensile strength of fibres compared with other commonly used natural fibres.*

fibres and fibres from across the globe have been compared. **Figures** 7 and **8** show how this is done. The table lists all of the governing mechanical characteristics of the most regularly used fibres. **Figure** 7 shows that palm fibres have the greatest tensile strength value when compared to lemon and mustard fibres.

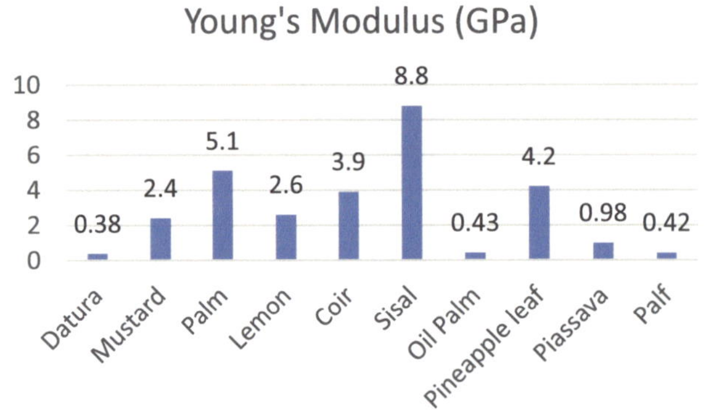


**Figure 8.**
*Young's modulus of fibres compared with other commonly used natural fibres.*

**Figure 8** shows a comparison of the tensile modulus of the most regularly used fibres. Except for sisal, it is obvious that Jharkhand fibres have a higher modulus value than the other fibres. The Young's modulus of mustard and lemon fibres, for example, is higher than that of palf, oil palm, and piassava fibres. As a consequence, Jharkhand's agro-based waste fibres have a better chance of becoming biomaterials for industrial use.

### 3.2 Thermographic analysis (TGA)

Agricultural waste produced all sorts of fibres studied in the literature. Thermal procedures have looked at the influence of heat on their characteristic of weight loss before combustion to see whether this material is acceptable for future manufacturing processes of environmentally friendly products. TGA testing was also used to track the physical vagaries of raising the temperature rate at a regular pace. This approach was also used to regulate the chemical characteristics of the fibres. Water losses must be determined throughout the breakdown process particularly for agricultural waste fibres [15]. The TGA research revealed that palm, datura, lemon, and mustard are all thermally stable, with minor/negligible behaviour when it comes to mass losses.

Thermal stability was measured with many specimens in the temperature range of 31–300°C and 500°C at intervals of 10°C/min heating rate and found to be stable. As the temperature rose, people began to lose weight as a result of the heat. It was discovered that the first percentage loss in mass is 7.2% up to 140°C. The cause for this is the evaporation of water, as well as the presence of lignin. Depending on the fibre type, the moisture content varies. By raising the temperature over this point, the other elements of fibres, such as hemicellulose and lignin type cellulose, were harmed. In this study, the degradation degree of lemon fibres was shown to be divided into three phases. The second and third levels begin from 190–240°C and 240–300°C, respectively. At these two levels, the percentage of deterioration was 5.17% of mass and 14.82% of weight, respectively.

The entire analysis of the fibres addressed in this paper is shown in **Figure 9**. This diagram depicts the thermal behaviour of fibres in terms of stability and degradation. In the experiments, lemon and datura fibres were shown to be more stable for the initial temperature range of 140–240°C. At a temperature of 240°C, datura and lemon fibres are steadier in percentage weight loss than palm and mustard fibres at 140°C. Except for lemon fibres, all other fibre materials studied had a weight decrease of greater

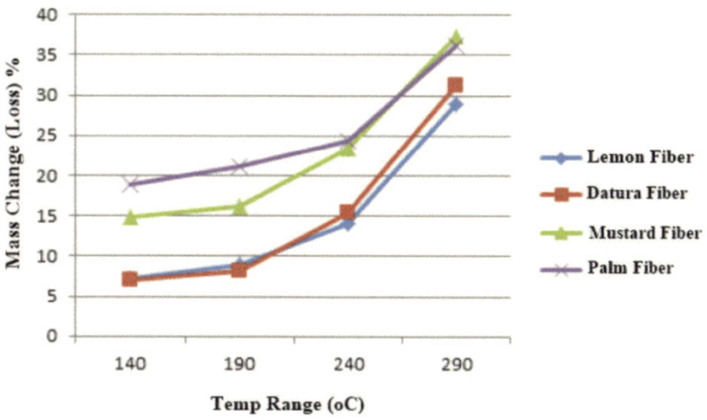


**Figure 9.**
*Thermal stability of fibres compared.*

than 30%. This fibre material's stable feature is better suited for biomaterials based on polymers for industrial applications requiring a greater thermal stability value.

As a result of the experiments, it has been theoretically shown that fibres from Jharkhand agricultural wastes are more appropriate for industrial applications owing to their better mechanical and electrical characteristics. Insulation in door panels, covering door racks, window frames, home railings, furniture industries, paper, textiles, insulated electronics, circuit boards, terminals, switches, and dielectrics are just a few examples.

**Figure 10** shows that, when it comes to divergence to percentage mass loss, Jharkhand fibres outperform other fibre types. Datura and lemon have been demonstrated to have a lower weight reduction percentage than bagasse, banana, bamboo, pineapple leaf, and phoenix SP. It demonstrates Jharkhand's increased stability and enhanced opportunities for green product manufacturing.

### 3.3 The pullout test

The pull-out technique is one of the most important ways to confirm that lignocellulosic fibres and the different polymers investigated in this study are compatible. By assuring the interfacial bonding capacity, this approach determines the maximum load applied to the fibre up to which it can withstand the load limit. This approach may be used to determine if agricultural waste fibres and polymer components are

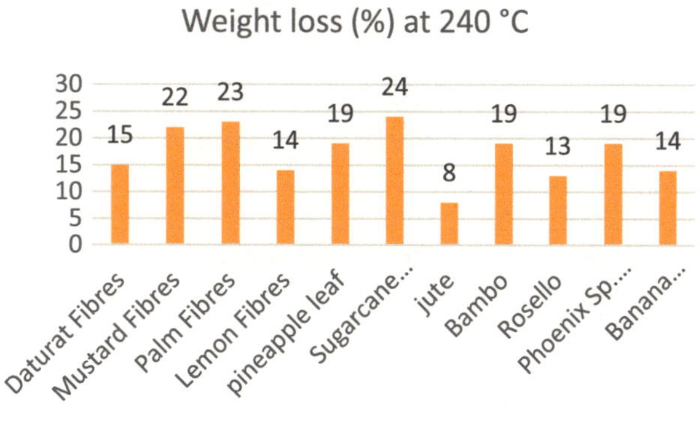


**Figure 10.**
*Weight loss (%) of fibres compared with other commonly used natural fibres.*

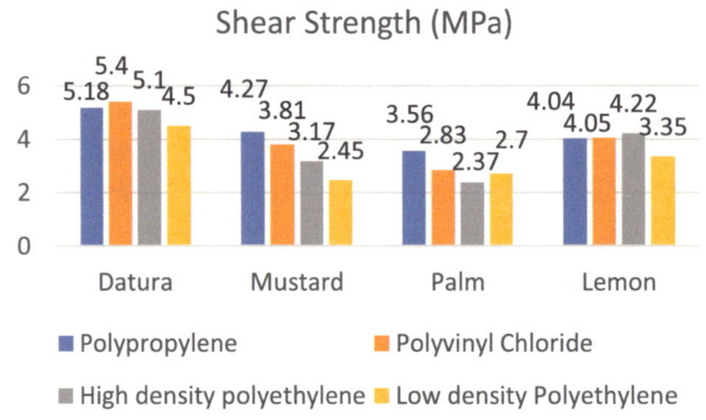


**Figure 11.**
*Shear strength of fibres compared through use as polymers.*

compatible. It was decided to prepare samples for the test hot mixing extruding technique. Polypropylene, polyethylene of high density, polyethylene of low density, and epoxy were among the samples bought and prepared for the pull-out test [9, 16, 17].

**Figure 11** depicts the shear strength of the different polymer interfaces. The graph shows that datura fibre has the best interfacial bonding, and that PVC polymer has a shear strength of 5.3 MPa. Datura's most important attribute is its coarser surface compared to other polymers, which allows for sticky properties. Because of their flat surfaces, mustard and palms have reduced interfacial bonding.

## 4. Conclusion

This research successfully achieves the interfacial properties of Jharkhand lignocellulosic fibres. Jharkhand fibres' thermal and mechanical properties were also examined in laboratory tests. A side-by-by-side comparison of foreign and Jharkhand fibres was carried out in order to assess their inherent capabilities and usefulness. Because of its increased mechanical strength, thermal stability, and strong adhesive forces at surfaces, the Jharkhand lignocellulosic was determined to be more suited. For biomaterials applications, Jharkhand's fibres have also been shown to be better compatible with a variety of polymers. Palm-type fibre materials have higher mechanical strength, elongation to break, and tensile strength than mustard fibres. The thermal stability of the abovementioned fibres is determined to be better in the case of lemon fibres, while datura fibres are best at 240°C and 290°C owing to their mass to loss percentage.

Lemon and datura fibres are the most ideal for thermal properties when compared to pineapple leaf, bamboo, roselle, bagasse, and phoenix SP, since they have a lower weight ratio at 240°C. Furthermore, after a thorough examination of fibre materials, it is possible to conclude that, due to their cheap cost and environmental friendliness, widespread manufacturing of these green goods might be boosted for developing nations. With the Jharkhand fibre's better interfacial bonding, sustainable businesses may now produce green products. PVC and datura fibres were found to be the best fibres for shear loads with all types of polymers in this paper. Doors, textiles, packaging and papers, furniture, window frames and electrical applications such as insulators, circuit boards and dielectrics are just a few of the many uses for Jharkhand waste agro fibres.

## Acknowledgements

The authors are extremely grateful to the entire fraternity at the National Institue of Technology, Durgapur, India, for their constant support and guidance.

## Data availability statement

All data, models, and code generated or used during the study appear in the submitted article.

## Author details

Dheeraj Kumar*, Nadeem Faisal, Ramit Choudhury and Swarup S. Deshmukh
Department of Mechanical Engineering, National Institute of Technology, Durgapur, India

*Address all correspondence to: dknitdgp1@gmail.com

## References

[1] Al-Oqla FM. Investigating the mechanical performance deterioration of Mediterranean cellulosic cypress and pine/polyethylene composites. Cellulose. 2017;**24**:2523-2530

[2] Al-Oqla FM, El-Shekeil YA. Investigating and predicting the performance deteriorations and trends of polyurethane bio-composites for more realistic sustainable design possibilities. Journal of Cleaner Production. 2019;**222**: 865-870

[3] Asim M, Jawaid M, Abdan K, Ishak M. The effect of silane treated fibre loading on mechanical properties of pineapple leaf/kenaf fibre filler phenolic composites. Journal of Polymers and the Environment. 2018;**26**:1520

[4] Almagableh A, Al-Oqla FM, Omari MA. Predicting the effect of nano-structural parameters on the elastic properties of carbon nanotube-polymeric based composites. International Journal of Performability Engineering. 2017;**13**:73

[5] Al-Oqla FM, Sapuan SM, Fares O. Electrical–based applications of natural fiber vinyl polymer composites. In: Natural Fibre Reinforced Vinyl Ester and Vinyl Polymer Composites. Woodhead Publishing; 2018. pp. 349-367

[6] Al-Oqla FM, Salit MS. Materials Selection for Natural Fiber Composites. Cambridge, USA: Woodhead Publishing, Elsevier; 2017

[7] Sanjay MR, Madhu P, Jawaid M, Senthamaraikannan P, Senthil S, Pradeep S. Characterization and properties of natural fiber polymer composites: A comprehensive review. Journal of Cleaner Production. 2018;**172**:566-581

[8] Väisänen T, Das O, Tomppo L. A review on new bio-based constituents for natural fiber-polymer composites. Journal of Cleaner Production. 2017;**149**:582-596

[9] Kumar D, Faisal N, Layek A, Priyadarshi G. Enhancement of mechanical properties of carbon and flax fibre hybrid composites for engineering applications. In: AIP Conference Proceedings. Vol. 2341, No. 1. AIP Publishing LLC; 2021. p. 040032. DOI: 10.1063/5.0050313

[10] Otto GP, Moisés MP, Carvalho G, Rinaldi AW, Garcia JC, Radovanovic E, et al. Mechanical properties of a polyurethane hybrid composite with natural lignocellulosic fibers. Composites Part B: Engineering. 2017;**110**:459-465

[11] Al-Oqla FM, Hayajneh MT, Fares O. Investigating the mechanical thermal and polymer interfacial characteristics of Jordanian lignocellulosic fibers to demonstrate their capabilities for sustainable green materials. Journal of Cleaner Production. 2019;**241**:118256

[12] Shi S, Yang C, Nie M. Enhanced interfacial strength of natural fiber/polypropylene composite with mechanical-interlocking interface. ACS Sustainable Chemistry & Engineering. 2017;**5**(11):10413-10420

[13] Hoque MB, Hossain MS, Nahid AM, Bari S, Khan RA. Fabrication and characterization of pineapple fiber-reinforced polypropylene based composites. Nano Hybrids Compos. 2018;**21**;31-42. DOI: 10.4028/www.scientific.net/NHC.21.31

[14] Sapuan SM, Ismail H, Zainudin ES. Natural Fiber Reinforced Vinyl Ester and Vinyl Polymer Composites: Development,

Characterization and Applications. Woodhead Publishing Series in Composites Science and Engineering Elsevier; 2018

[15] Lila MK, Singhal A, Banwait SS, Singh I. A recyclability study of bagasse fiber reinforced polypropylene composites. Polymer Degradation and Stability. 2018;**152**:272-279

[16] Milosevic M, Stoof D, Pickering KL. Characterizing the mechanical properties of fused deposition modelling natural fiber recycled polypropylene composites. Journal of Composites Science. 2017;**1**(1):7

[17] Faisal N, Pandey MK, Chauhan T, Verma G, Farrukh M, Kumar D. Experimental investigation on mechanical behavior of palm based natural fibre reinforced polymer composites. AIJR Abstracts. 2021:161. ISBN:978-81-947843-2-6; DOI: 10.21467/abstracts.108

Chapter 4

# Recycled Synthetic Polymer-Based Electrospun Membranes for Filtering Applications

*Alena Opálková Šišková, Heba M. Abdallah, Smaher Mosad Elbayomi and Anita Eckstein Andicsová*

## Abstract

Synthetic polymers have been widely applied in various commercial and household applications owing to their fascinating properties of low-cost, lightweight, and processability. However, increasing population and living standards and rising demand for non-biodegradable polymers have led to the accumulation of plastic pollution resulting in the current environmental crisis. Current waste management methods such as landfilling or incineration do not solve these environmental issues. On the other hand, recycling plastic waste is the most valuable strategy for dealing with waste as raw material for high-value products. One of such products is filter membranes. Polymer fiber membranes as masks in pandemics have been one of the most sought-after products in recent years. Some types of plastic waste became a material source for the development of filter materials, which could contribute to the protection of human health. Utilizing the simple, cheap, and industrially available technological solution is also needed. Given the number of advantages, electrospinning is such a beneficial solution. The electrospun polymer waste-based membranes show excellent filtration performance and can carry many other functionalities. Therefore, this review article presents a brief overview of electrospun nanofibrous membranes based on synthetic plastic waste and summarizes the filtration performance of such membranes. This review will discuss the future perspectives of electrospun membranes as well.

**Keywords:** plastic waste, recycling, nanofibers, filtration, waste management

## 1. Introduction

A linear economy system based on taking, making, using, disposing, and polluting is responsible for millions of tons of waste that threaten people's lives, while its littering and leakage in the environment cause negative impacts on land and sea life and also affect air quality [1, 2]. In 2018, an estimated 359 million tons of plastic (synthetic polymers) were produced globally. Production is still increasing due to rising living standards and the growth of the population on the Earth. The problems of a linear, resource-to-waste economy and environmental issues are becoming more acute. More than 29 million tons/annually of plastics are collected in Europe. Due to the low biodegradability of synthetic plastic waste, they need to be disposed of at the end of its service life.

Only 32.5% of plastics are recycled, 24.9% are landfilled, and 42.6% are incinerated. The traditional recycling methods aim to eliminate plastic waste rather than encourage its beneficial, value-added reuse. Landfilled plastics create an environmental burden. For example, all types of synthetic plastics, including polyester, polyamides, acrylic, and olefins, have been found in oceans, rivers, and even water treatment plants [3, 4]. When incinerated, the released toxic gasses create more environmental issues [5]. Air pollution has consistently been one of the most substantial ecological issues with serious consequences for human health regarding industrial growth.

On the contrary, reusing or recycling plastic waste into valuable products aligns well with the circular economy concept; plastics are recirculated by extending product life beyond one cycle.

The pandemic situation since 2019 caused the polymer fibrous membranes as masks against COVID 19 to be the most demanded products ever. Persistence of the dire situation and occurrence of other coronavirus mutations or other respiratory diseases have apparently started a new wave of filter media development. At a time when energy costs are rising, and materials, oil, and gas fields are being emptied or become unavailable, current plastic waste needs to be considered as a possible source of material in the production of high value-added products and filter media to protect human health, and the environment has almost priceless value.

In this respect, electrospinning is simple and industrially available technology for the production of randomly placed nanofibers into the nonwoven membrane suitable for filtration. The principle of electrospinning is based on the electrical discharge from liquid surfaces. The electrostatic attraction of liquid was observed in 1600 by Wiliam Gilbert. However, the first electrospinning patent was filled by J. F. Cooley in 1900. For over 120 years, electrospinning has evolved so much that it has become attractive to thousands of scientists around the world, and electrospinning devices are known in various designs. Since 1995, the number of publications about electrospinning has been increasing exponentially every year [6]. Semi-industrial and industrial equipment is being developed and sold in many countries. However, electrospinning enters the industry only very slowly, and to the best of the authors' knowledge, there is no industrial processing of plastic waste using electrospinning on an industrial scale. Electrospinning deserves a broader discussion, and therefore, Section 2 is addressed in this review.

One of the many advantages of electrospinning is that more than 200 polymers can be processed. Several authors in their studies have demonstrated the suitability of electrospinning in the processing of waste plastics. These are mainly wastes such as poly(ethylene terephthalate) (PET), polystyrene (PS) or expanded polystyrene (EPS), polyamides (PA), and polyurethanes (PUR), or thermoplastic polyurethane (TPU), but also waste natural polymers such as lignin and their mixtures. These membranes have already been studied for air or water filtration, water treatment, and cleaning from pharmaceutical, textile dyes, or oil. Such membranes show the filtration efficiency closed to 100%. The filtration efficiency depends on many factors, including the basis weight and thickness of the membrane or fiber diameters. Other functionalities such as antibacterial activity or wettability could be given to the membranes by additives, e.g., drugs, agents, nanoparticles, other polymers, etc. Despite the exciting possibilities in the field of filtration offered by recycling plastic waste by employing electrospinning, there are still not many articles dealing with this topic. There are still possibilities and perspectives that need to be further explored for expanded exploration of potential industrial applications.

This review article presents a brief overview of synthetic plastic waste-based electrospun nanofibrous membranes for filtration. The filtration performance of such membranes is summarized. The review also consists of the future perspectives of such electrospun membranes, given the composition of the plastics coming into the recycling process.

## 2. Electrospinning

Electrospinning is a versatile and straightforward technique to produce fine fibers on the scale of 20–1000 nm from polymer solutions or melts. The fibers are spun in a high electric field. Under a powerful electric field force, the polymer solution or melt overcomes its surface tension and viscoelastic force to generate a charge, the mutual exclusion between charges, and the opposite charge electrode's compression to the surface charge, directly creating a force that is opposite to the surface tension. When the electric field strength exceeds a specific critical value, the polymer solution or melt will overcome the droplet's surface tension and form a jet. The apex (Taylor cone) is formed at the spinneret (syringe, pipette, wire, rotating cylinder) from polymer drop due to the high electric field. The polymer jet is drawn, elongated, and stretched from the apex into a straight line to a certain distance; it is sprayed along a spiral path. As the solvent volatilizes, fibers solidify and are finally deposited on the grounded collector [1, 2, 7]. The principle of the electrospinning process is displayed in **Figure 1**.

The fiber morphology depends on parameters that can be divided roughly into several basic groups: polymer properties (molar mass, solubility) [8], solvent properties (boiling point, volatility, dielectric constant) [9], polymer solution properties (concentration, viscosity, conductivity) [10], the process parameters (applied voltage, top of the needle to collector distance, flow rate, needle diameter or cylinder rate) [11], and environmental parameters (humidity and temperature) [12] or properties of additives as plasticizers, drugs, dyes, etc. [13–15]. The effects of some

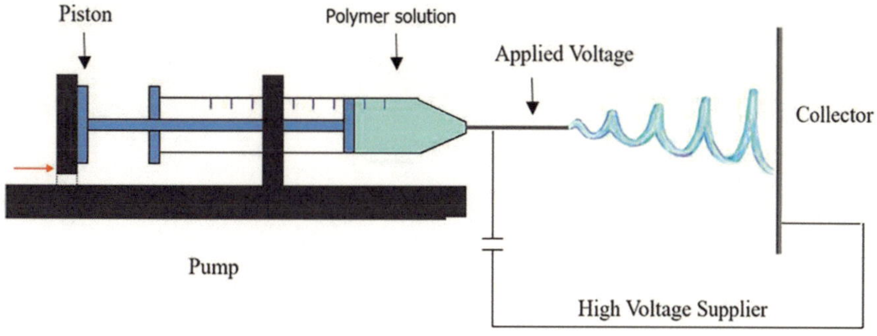


**Figure 1.**
*Schematic representation of electrospinning principle. From archive of authors.*

parameters are shown in **Figure 2**. Recently, various electrospinning technique modifications have been developed, such as needless electrospinning, multiple-jet electrospinning, coaxial electrospinning, bubble electrospinning, cylindrical porous hollow tube electrospinning, electro-blowing, magnetic-field-assisted electrospinning, and charge injection electrospinning [16]. Electrospinning has several advantages: it is a low-cost and commercially available technique for industrial production; it is a controllable process with high production efficiency [17]; electrospinning can be used for more than 200 different polymers, including synthetic as well as natural polymers [18, 19]; fibrous layers can be deposited onto a variety of collectors including metal plates, parallel electrodes, rotating-type disks or mandrels, knife-edge collectors, etc. [20]; fibers can be functionalized before, during, and after spinning [2, 21]. Electrospun materials could be used in wide application range such as packaging [22], wound healing [23], tissue engineering [24], drug-releasing systems [25], membranes [26], sensors [27], batteries [28], solar cells [29], supercapacitors [30], catalysts [31], environmental remediation [32], protecting clothing [33], and in many others. Due to all the mentioned advantages, electrospinning is the primary choice for fabricating nanomaterials.

Electrospinning offers the opportunity to reuse polymeric waste, turning plastic waste into higher-value fibrous polymer products from virgin as well as post-industrial or post-consumer polymers. The randomly placed ultrafine fibers in the electrospun membranes are attractive as filtration materials. The high surface area to volume ratios, nanoporosity, vapor permeability, and good mechanical properties of nanofibrous membranes predestined them for air or water filtration or even personal protection against fine dirt and microbes with smaller dimensions than 100 nm or volatile organic compounds [14].

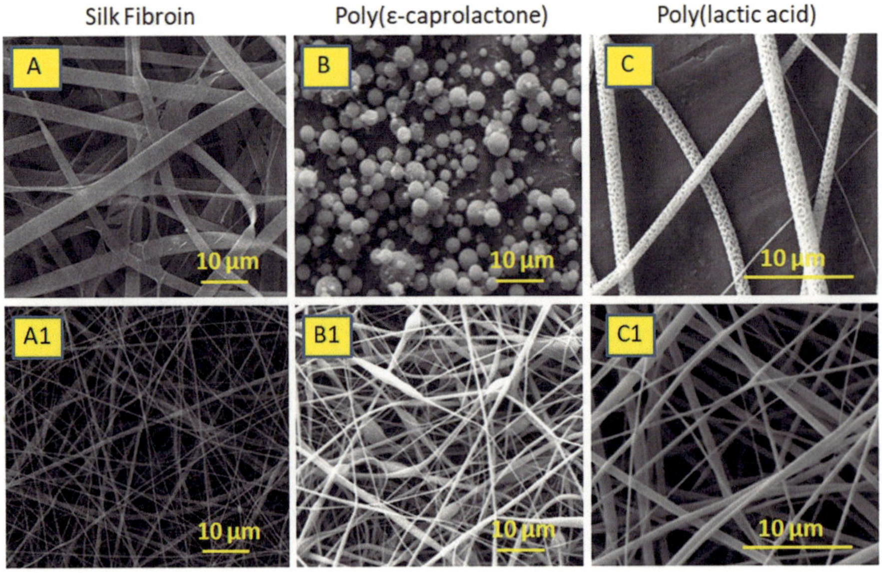


**Figure 2.**
*Effect of some electrospinning parameters on fibrous morphology. Here are represented: (A) silk fibroin (SF) with concentration solution 9 wt.%, (A1) with concentration 7 wt.%, other parameters were kept constant; (B) poly (ε-caprolactone) (PCL) with applied voltage 8 kV, (B1) with applied voltage 12 kV, other parameters were kept constant; (C) poly(lactic acid)(PLA) heated collector on 60°C, (C1) PLA ambient temperature of the collector, other parameters were kept constant. Images are from the archive of authors.*

## 3. Polymeric waste for filtration

Polymeric materials are inexpensive, lightweight, and durable materials processed into various products that are used in a wide range of consumer and industrial applications. Due to their benefits to our society and economy, their production has increased significantly [14].

Especially synthetic polymers such as polyethylene terephthalate (PET), polystyrene (PS), and polyamide (PA) have been widely applied in various applications since the 1940s thanks to the properties related to the processability and corrosion resistance [1].

PET is a low-cost, thermoplastic polyester with 18.3 MT, about 5% of global plastic production. Roughly 30% of that amount of PET is used in the food industry, including single-use plastic, automotive, electrical, electronics, and textile. PS or EPS, known as styrofoam, is used for insulation or packaging materials and accounts for another 5% of total annual plastics production. Polyamides (PAs) are semicrystalline polymers with high chemical resistance and good thermal and mechanical properties [2]. They are usually produced in the form of fibers for use in fashion, automotive, electrical, electronic, construction, packaging, and other industries. Currently, PAs are made on an annual worldwide multimillion-ton scale, and the production is estimated to be continuously growing [34]. Polyamides represent about 2% of total plastic production.

All the mentioned polymers represent an important proportion of plastic waste. Due to the low biodegradation of such plastic waste and increasing consumption created by a rising living standard and a growing world population, the problems with plastic waste and environmental issues are becoming more acute. Moreover, the increased use of mentioned plastics, primarily for packaging and textile, has heightened the importance of managing this material's end of life.

Due to the good processability of PET, PA, as well as PS, was processed by electrospinning and the conditions of production electrospun membranes have been already extensively described in the literature [35–39]. The randomly arranged ultrafine fibers in the electrospun membranes, the high surface area to volume ratios, nanoporosity, good mechanical properties, and vapor permeability of such membranes predestined them for filtration membranes and especially for personal protection as face masks against very fine dirt and bacteria, but also viruses with dimensions of about 100 nm [40, 41]. In the following subsections, the filtration performances of polymer waste-based electrospun membranes in air filtration and water treatment will be discussed.

### 3.1 Air filtration

The pandemic situation since 2019 caused the polymer fibrous membranes as masks against COVID 19 to be the most demanded products ever. Persistence of the dire situation and occurrence of other coronavirus mutations or other respiratory diseases have apparently started a new wave of filter media development. The new technologies are considered, but it is also necessary to consider the sources of materials. In this respect, every alternative source of materials should be seriously considered, and waste is perceived as a significant and low-cost source of the material.

Rajak et al. developed an air filter by electrospinning EPS waste of electronic appliances packaging [42]. The filtration efficiency for fibers with a diameter of about 3500 nm was 70.4%; however, the lower the fiber diameter, the higher the filtration efficiency. The fibers with a diameter of about 314 nm achieved an efficiency of almost 100%.

Most electrospun recycled PET fibrous membranes were applied for filtration [1]. Tough fibrous membranes for smoke filtration were developed from recycled PET bottles by Strain et al. The diameter of the fibers is about 410 nm, the large surface specific area of 7.07 $m^2.g^{-1}$ and affinity of the rPET mats for airborne hydrocarbons open the way for the applied use in a range of industrial filters [43]. Nosko et al. found that the electrospun recycled PET fibers from the bottle, with an average diameter of around 95 nm, show the highest filtration efficiency of 99.95%, compared with the cotton fabric and melt blown polypropylene nonwoven, of particles with a size of 120 nm. The membrane exhibits good vapor permeability but low breathability [44].

Bonfim et al. investigated the PET bottles waste membrane with average diameter fibers of about 1290 nm. It was shown that the filtration efficiency was 98.4%, with a pressure drop of 212 Pa [45]. Breathing comfort is generally correlated with a pressure drop [46]. Then, by investigating the effect of morphological characteristics on filtration performance, Bonfim et al. obtained a filter with an average diameter of the fibers of about 650 nm that showed a capture of 99% of nanoparticles, the low pressure drop of 19.4 Pa with outstanding mechanical properties [47].

Unlike PET, silk fibroin is a biopolymer obtained from natural silk cocoons. It is favorable due to its biocompatibility, minimal inflammatory reactions, water vapor, and oxygen permeability. Šišková et al. [48] recycled PET from domestic plastic waste and blended it with silk fibroin. In this study, the basis weight of the membrane plays a key role. With the increasing basis weight, the filtration efficiency increases, and the most effective membranes are comparable to the efficiency of HEPA filters according to the European Standard EN1822. Due to the high pressure drop, the membranes were not suitable for personal protection as filtration masks or respirators. However, with the decrease of the basis weight, the pressure drop decreased. Such membrane efficiency was comparable to the FFP1 class facial mask according to the European standard EN149. The tested membranes rPET/SF has been antibacterial against *Escherichia coli* (*E. coli*) and *Staphylococcus aureus* (*S. aureus*) and biocompatible with HaCaT cells spontaneously transformed aneuploid immortal keratinocyte from adult human skin.

Pleva et al. investigated the electrospun recycled polyamide (PA) from the women stocking in filtration. The filtration efficiency of rPA and rPA/MAG (monoacylglycerol) was 98.49 and 99.87% of particles with sizes about 600 nm; 92.70 and 96.10% of particles with sizes about 100 nm the pressure drop about 129 and 189 Pa, respectively. Moreover, the rPA/MAG membrane demonstrated antibacterial activity against *S. aureus* and antifouling activity against *E. coli* and *S. aureus* [14].

In **Table 1**, the filtration efficiencies of electrospun membranes from selected synthetic virgin and recycled polymers are compared.

The data in **Table 1** encourage the utilizing plastic waste for filtration.

### 3.2 Water treatment

Industrial wastewater discharges and oil spills have precipitated significant damage to water resources in recent years. Water pollution not only causes an imbalance in the ecological environment but also has a direct impact on human health. Scientists urgently need to develop effective wastewater purification technology in order to address this pressing problem. Several previously identified techniques for wastewater treatment, such as adsorption, filtration, centrifugation, catalysis, biological

| Polymer | Fiber diameter [nm] | Basis weight [g.m$^{-2}$] | PM Size [nm] | E [%] | ΔP [Pa] | Ref. |
|---|---|---|---|---|---|---|
| **Virgin polymer** | | | | | | |
| Polyamide 66 (PA66) | 60 | 0.46 | 300 | 90.9 | 69 | [35] |
| Polyamide 6 (PA6) | 120 | 0.63 | 300–500 | 99.9 | 95 | [38] |
| Polyurethane (PU) | 120 | 0.4–0.9 | 20–400 | 99.7 | 96–190 | [36] |
| Poly(ethylene terephthalate)/ thermoplastic polyurethane (PET/ TPU) | 395 | NA/ the thickness is given 35 μm | 50–500 | 83.6 | 28.9 | [37] |
| **Recycled polymer** | | | | | | |
| Poly(ethylene terephthalate) (PET) | 95 | 14 | 120 | 98.3 | NA | [44] |
| Poly(ethylene terephthalate) (PET) | 230 | 12 | 120 | 99.9 | 414 | [48] |
| Poly(ethylene terephthalate) (PET) | 230 | 3 | 120 | 96.0 | 123 | [48] |
| Polyamide 66 (PA66) | 640 | 8.7 | 300 | 97.7 | 129 | [14] |
| Poly(ethylene terephthalate) (PET) | 1290 | The thickness was given 342–392 | up to 100 | 98.4 | 212 | [45] |
| Poly(ethylene terephthalate) (PET) | 650 | The thickness was given 186–392 | up to 100 | 99.0 | 19.4 | [45] |
| PET | 1240–3180 | The thickness is given over 100 μm | 500–10,000 | 98.2 | 36 | [38] |

**Table 1.**
*Filtration performance of electrospun virgin and recycled polymers.*

treatment, and electro-coalescence, have been commonly implemented [49–54]. Whereas high energy consumption, secondary pollution, and unrepeatable usage still appear inescapable.

At the beginning of twenty-first century, nanotechnology has become one of the most competitive scientific fields. It has been widely regarded as the most significant technology of the new industrial revolutions [55, 56]. As a result, nanofibers are predicted to be appropriate materials for wastewater treatment. The following subsections describe and illustrate (**Figure 3**) the most recent and significant applications of nanofibers from recycled plastics in wastewater treatment to remove various pollutants, especially in oil/water filtration and removal of pharmaceutics.

### *3.2.1 Oil/water filtration*

Metallurgy, transportation, food processing, petrochemicals, petroleum, natural gas, and pharmaceuticals usually produce a significant amount of oily wastewater daily [57]. Oily compounds are insoluble in water and can be toxic. Therefore,

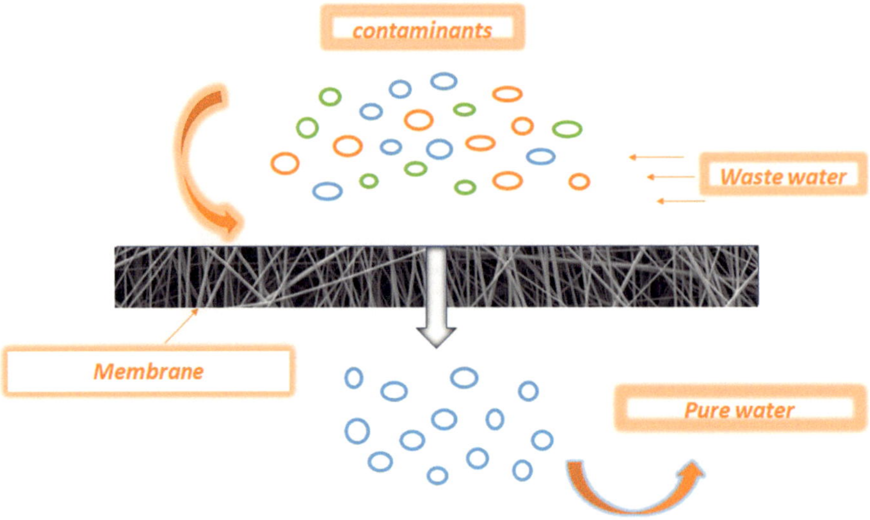


**Figure 3.**
*Schematic representation of the liquid filtration process. Figure is from the archive of authors.*

approaches such as coagulation, flocculation, air flotation, chemical degradation, and membrane filtration have been used to separate oil/water [58, 59]. However, the general oil/water separation method has low separation efficiency, high energy cost, and complicated operation. In correlation, membrane separation generally exhibits superior separation performance, lower energy requirements, and more straightforward preparation methods. The nanofiber membranes' performance is particularly outstanding [60].

In response to the growing demand for cost-effective and sustainable methods of producing fibrous membranes, numerous studies have been conducted to determine the efficacy of using waste polymers in separating oil and water. For instance, Liu et al. coated a stainless mesh with waste cigarette filters to create an electrospun membrane capable of separating oil–water mixtures and emulsions [61]. Sow et al. also used waste polystyrene (PS) to fabricate superoleophilic fiber-coated membranes for oil recovery through blow spinning. These membranes demonstrated a separation efficiency of up to 97% [62]. Regrettably, recycled PET (rPET) bottles are frequently utilized to produce low-cost products, resulting in relatively low market profits. Utilizing rPET as a starting material for the fabrication of fibrous membranes for oil–water separation could result in cost savings and environmental benefits. Several studies on the use of rPET in fibrous filters have already been performed. Zander et al. recently designed rPET nanofibers for filtering particles dispersed in water with diameters ranging from 30 to 2000 nm [63]. Additionally, the oil–water separation performance of functionalized rPET was evaluated previously, with separation efficiencies exceeding 98.5%. Doan et al. produced flexible fibrous membranes with a high capacity for oil–water separation based on recycled rPET via electrospinning. Dip coating the rPET membrane with polydimethylsiloxane (PDMS) improved the rPET fibrous membrane's water intrusion pressure and anti-water-fouling property [64]. Nanofibrous membranes have recently been developed to control oil and water separation. Smart materials with the ability to respond to temperature [65], pH [66], light [67], ions [68], electric fields [69], and prewetting are emerging candidates for on-demand oil–water separation. Prewetting appears to be the most promising strategy due to its facile fabrication and operation [70, 71]. The prewetting membrane should have amphiphilic properties, which means it should be superoleophobic underwater and superhydrophobic underoil. Baggio et al. developed a nanofibrous membrane from the blend of rPET and chitosan, which have amphiphilic properties because of the presence of the hydrophobic groups of rPET and the hydrophilic groups of chitosan. The resulting filter exhibited high antifouling properties and separation efficiency [72].

#### *3.2.2 Removal of pharmaceuticals and dyes*

A literature review shows that only a few studies on the preparation membranes from recycled plastic waste have been conducted for the removal of organic compounds.

Zander et al. synthesized nanofibrous microfiltration membranes from recycled PET and evaluated their filtration performance with latex beads [63]. Xu et al. utilized recycled PET bottles to develop electrospun nanofibrous membranes, modified them through fluorination, and applied them in membrane distillation [73]. Arahman et al. fabricated rPET/polyvinylpyrrolidone (PVP) ultrafiltration membranes for humic acid separation from water [74]. Pulido et al. prepared rPET/PEG ultrafiltration membranes to filtrate PEG/water and PEG/dimethylformamide solutions [75]. Waste PET bottles were used to fabricate PET ultrafiltration membranes via Kusumocahyo et al. [76]. They utilized phenol as the solvent, PEG as the additive, water-ethanol, water-n-propanol, and water-n-butanol as the non-solvent bath. Using waste PET bottles as a starting point, Kusumocahyo et al. confirmed the fabrication and characterization of PET ultrafiltration membranes, PEG 400 as the additive, and water-ethanol as the non-solvent. Higher porosity and smaller pore sizes were achieved by increasing the number of additives. Utilizing recycled PET will unquestionably minimize waste synthetic polymer production [77]. Kiani et al. demonstrated an environmentally friendly nanofiltration membrane with high performance synthesized through recycled PET bottles. Xanthan gum (XA) was introduced to the membranes as a hydrophilic additive during membrane fabrication, using water and methanol as coagulation baths. The produced membranes were used in the nanofiltration of an aqueous solution incorporating diltiazem [78], a calcium channel blocker used to treat hypertension. It frequently causes edema, headaches, and dizziness [79].

Preparation and application of electrospun membranes composed of lignin extracted from palm fronds and banana bunches as biomass sources and PET from bottle waste were studied by Attia et al. [80]. The electrospun fibers with smooth morphology for the adsorption of methylene blue dye from the water were investigated. Fibers with 1800 ± 500 nm diameter were obtained. The iodine-treated membrane was carbonized at 500°C, and the achieved carbon content did not exceed 62 wt %. The reported adsorption capacity was about 9 mg of MB/g CFs of capacity. The work describes the potential of combining biomass waste as a lignin source with sustainable waste (PET) to produce fibers as a core material that can be used for environmental remediation. Similar work was described by Yasin et al. [81], where the electrospinning made PET nanofibers from the wasted PET polymer containing CuO nanoparticles. The nanoparticle was cross-linked on the surface of the electrospun PET nanofibers to enhance photodegradation of the methylene blue dye and adsorption of the degradation products. Thus produced eco-friendly, cheap, sustainable, and effective nanocomposite has high efficiency at the dye removal (above 99% in 30 min). The composite of the base of $TiO_2$/styrofoam membranes prepared using electrospinning was developed by Rajak et al. [82] to purify water from harmful organic compounds such as bacteria, viruses, or even textile dyes. Styrofoam from waste was used for the study, and $TiO_2$ as a semiconductor catalyst was used in the environmentally friendly photocatalysis process. Because $TiO_2$ has high photocatalytic activity, the fibrous composites showed photocatalytic activities too.

## 4. Future perspectives

Future perspectives could be concluded from the following aspects:

In many cases, plastic waste is a composite, so it is necessary to combine electrospinning with other technology that will ensure a high yield of polymer, which could be subsequently processed into nano/micro-fibers;

I. Study of the effect of additives in the wasted plastics coming to the recycling process on the processing parameters and resulted properties.

II. Vice versa, study of advantageous additives, nanofillers, or mixed plastic waste to make the electrospun composites with special functions for special applications, for example, with antibacterial or antifouling activity;

III. The morphology and properties of electrospun membranes are sensitive to electrospinning parameters; therefore, the deep research on the optimization of electrospun processing parameters for different types of plastic waste to produce high-value products;

IV. Further study of natural or benign solvents for solution electrospinning or practical design solutions for melt spinning. To realize "green" production;

V. Combination of electrospinning with other auxiliary devices to enhance the mechanical properties, for example, ultrasound sonication; or device to collect the evaporated solvent and its reuse.

VI. Intensive study on mass production of electrospun membranes from plastic waste to reach commercialization and standardization.

## 5. Conclusions

Rising worldwide attention on polymeric waste leads to the founding of new ways for their environmentally safe utilization. A European strategy for polymeric waste use focuses on designing and producing polymeric materials, which can be reused or recycled. For this reason, there is an urgent need to promote the utilization of polymeric waste after its end of life. Upcycling plastic waste into high-value products for various applications is a sustainable solution with a promising future. As it is shown in this study, plastic waste is an excellent material source for the production of filter materials and the protection of human health and the environment.

Moreover, electrospinning is undoubtedly a superior method to process plastic waste into functional nano/micro-fibers due to the randomly placed ultrafine fibers, high surface area to volume ratios, nanoporosity, vapor permeability, and good mechanical properties. These properties predestined the nanofibrous membranes for air or water filtration or even personal protection against fine dirt and microbes with smaller dimensions than 100 nm or volatile organic compounds. In this review, reusing plastic waste, especially non-degradable polymers PET, PS, PUR, and PA, and the present advances in recycling plastic waste by electrospinning were briefly summarized. The filtration efficiency was from 80 to 99.9%, depending on the thickness and basis weight of the membranes. The average diameter of the fibers also plays a role in the efficiency. It must be noted that the pressure drop, which represents the ease of air passage when breathing, also depends on the basis weight and thickness. However, as the thickness increases, the filtration efficiency improves, but pressure drop increases, so the breathability is more complicated. Therefore, talking about using these materials in face masks or respirators, it is necessary to look for the golden mean in all the parameters affecting the parameters of the final product. Here was also shown that the membranes are a good alternative to the EPA and HEPA filters. The suitable additives such as carbon fibers, CuO, $TiO_2$ or other polymers as PVP, PEG, or Xanthan gum give the membranes functionalities used in air and water cleaning or remediation. The excellent availability of waste plastics and electrospinning-induced formation of tough fibrous mats pave the way for using recycled plastics in industrial air filters or membranes for water treatment.

## Acknowledgements

The authors acknowledge the support by the Scientific Grant Agency of the Ministry of Education, Science, Research and Sport of the Slovak Republic under project no. VEGA 2/0168/21 and no. VEGA 2/0143/22. The present work was also supported by the Slovak Research and Development Agency under contracts no. APVV-19-0338 and by project SAS-MOST JRP 2019/07. The authors would like to acknowledge networking support by the COST Actions CA17107, CONTEXT, European Network to connect research and innovation efforts on advanced Smart Textiles and CA20101, PRIORITY, Plastics monitoring detection remediation recovery.

## Conflict of interest

Authors declare that there is no conflict of interest.

## Author details

Alena Opálková Šišková[1,2*], Heba M. Abdallah[2,3], Smaher Mosad Elbayomi[4] and Anita Eckstein Andicsová[2]

1 Institute of Materials and Machine Mechanics, Slovak Academy of Sciences, Bratislava, Slovakia

2 Polymer Institute of Slovak Academy of Sciences, Bratislava, Slovakia

3 Polymers and Pigments Department, Chemical Industries Research Institute, National Research Center, Giza, Egypt

4 Department of Chemistry, Damietta University, Damietta, Egypt

*Address all correspondence to: alena.siskova@savba.sk

## References

[1] Li X, Peng Y, Deng Y, Ye F, Zhang C, Hu X, et al. Recycling and reutilizing polymer waste via electrospun micro/nanofibers: A review. Nanomaterials. 2022;**12**:1663. DOI: 10.3390/nano12101663

[2] Opálková Šišková A, Peer P, Eckstein Andicsova A, Jordanov I, Rychter P. Circulatory management of polymer waste: Recycling into fine fibers and their application. Materials. 2021;**14**:4694. DOI: 10.3390/ma14164694

[3] Novotna K, Cermakova L, Pivokonska L, Cajthaml T, Pivokonsky M. Microplastics in drinking water treatment – Current knowledge and research needs. Science of the Total Environment. 2019;**667**:730-740. DOI: 10.1016/j.scitotenv.2019.02.431

[4] Pivokonsky M, Cermakova L, Novotna K, Peer P, Cajthaml T, Janda V. Occurrence of microplastics in raw and treated drinking water. Science of the Total Environment. 2018;**643**:1644-1651. DOI: 10.1016/j.scitotenv.2018.08.102

[5] Verma R, Vinoda KS, Papireddy M, Gowda ANS. Toxic pollutants from plastic waste – A review. Procedia Environmental Sciences. 2016;**35**:701-708. DOI: 10.1016/j.proenv.2016.07.069

[6] Tucker N, Stanger JJ, Staiger MP, Razzaq H, Hofman K. The history of the science and technology of electrospinning from 1600 to 1995. Journal of Engineered Fibers and Fabrics. 2012;7:63-73

[7] Liu Y, Li K, Mohideen MM, Ramakrishna S. Melt Electrospinning: A Green Method to Produce Superfine Fibers. 1st ed. Amsterdam: Elsevier Inc.; 2019 202 p. ISBN: 978-0-12-816220-0

[8] Mwiiri FK, Daniels R. Influence of PVA molecular weight and concentration on electrospinnability of birch bark extract – Loaded nanofibrous scaffolds intended for enhanced wound healing. Molecules. 2020;**25**:4799. DOI: 10.3390/molecules25204799

[9] Guerrini LM, Oliveira MP, Stapait CC, Maric M, Santos AM, Demarquette NR. Evaluation of different solvents and solubility parameters on the morphology and diameter of electrospun pullulan nanofibers for curcumin entrapment. Carbohydrate Polymers. 2021;**251**:117127. DOI: 10.1016/j.carbpol.2020.117127

[10] Mirtič J, Balažic H, Zupančič Š, Kristl J. Effect of solution composition variables on electrospun alginate nanofibers: Response surface analysis. Polymers. 2019;**11**:692. DOI: 10.3390/polym11040692

[11] Feng SM, Liu XL, Qi J, Huang DL, Xiong ZC. Effect of electrospinning parameters on morphology of polydioxanone nanofibers. Materials Research Express. 2019;**6**:125330. DOI: 10.1088/2053-1591/ab58e0

[12] Nezarati RM, Eifert MB, Cosgriff-Hernandez E. Effect of humidity and solution viscosity on electrospun fiber morphology. Tissue Engineering. Part C, Methods. 2013;**19**:810-819. DOI: 10.1089/ten.tec.2012.0671

[13] Homaeigohar S, Boccaccini AR. Nature-derived and synthetic additives to poly(ε-caprolactone) nanofibrous systems for biomedicine; an updated overview. Frontiers in Chemistry. 2022;**9**:809676. DOI: 10.3389/fchem.2021.809676

[14] Opálková Šišková A, Pleva P, Hrůza J, Frajova J, Sedlaříková J, Peer P, et al.

Reuse of textile waste to production of the fibrous antibacterial membrane with filtration potential. Nanomaterials. 2022;**12**:50. DOI: 10.3390/nano12010050

[15] Koenig K, Hermanns S, Ellerkmann J, Saralidze K, Langensiepen F, Seide G. The effect of additives and process parameters on the pilot-scale manufacturing of polylactic acid sub-microfibers by melting electrospinning. Textile Research Journal. 2020;**90**:1948-1961. DOI: 10.1177/0040517520904019

[16] Alghoraibi I, Alomari S. Different methods for nanofiber design and fabrication. In: Bahroum A, Bechcelany M, Makhlouf A, editors. Handbook of Nanofibers. 1st ed. Cham: Springer; 2018. pp. 1-46. DOI: 10.1007/978-3-319-42789-8_11-2

[17] Ahmed FE, Lalia BS, Hashaikeh R. A review on electrospinning for membrane fabrication: Challenges and applications. Desalination. 2015;**356**:15-30. DOI: 10.1016/j.desal.2014.09.033

[18] Keshvardoostchokami M, Majidi SS, Huo P, Ramachandran R, Chen M, Liu B. Electrospun nanofibers of natural and synthetic polymers as artificial extracellular matrix for tissue engineering. Nanomaterials. 2021;**11**:21. DOI: 10.3390/nano11010021

[19] Opálková Šišková A, Kozma E, Opálek A, Kroneková Z, Kleinová A, Nagy Š, et al. Diclofenac embedded in silk fibroin fibers as a drug delivery system. Materials. 2020;**13**:3580. DOI: 10.3390/ma13163580

[20] Zander NE. Hierarchically structured electrospun fibers. Polymers. 2013;**5**:19-44. DOI: 10.3390/polym5010019

[21] Song Y, Li L, Zhao W, Qian Y, Dong L, Fang Y, et al. Surface modification of electrospun fibers with mechanical-growth factor for mitigating the foreign-body reaction. Bioactive Materials. 2021;**6**:2983-2998. DOI: 10.1016/j.bioactmat.2021.02.020

[22] Dehghani S, Rezaei K, Hamishehkar H, Oromiehe A. The effect of electrospun polylactic acid/chitosan nanofibers on the low density polyethylene/polylactic acid film as bilayer antibacterial active packaging films. Journal of Food Processing and Preservation. 2021:e15889. DOI: 10.1111/jfpp.15889

[23] Xiang J, Zhou L, Xie Y, Zhu Y, Xiao L, Chen Y, et al. Mesh-like electrospun membrane loaded with atorvastatin facilitates cutaneous wound healing by promoting the paracrine function of mesenchymal stem cells. Stem Cell Research & Therapy. 2022;**13**:190. DOI: 10.1186/s13287-022-02865-5

[24] Kang Y, Wang C, Quao Y, Gu J, Zhang H, Peijs T, et al. Tissue-engineered trachea consisting of electrospun patterned sc-PLA/GO-g-IL fibrous membranes with antibacterial property and 3D-printed skeletons with elasticity. Biomacromolecules. 2019;**20**(4):1765-1776. DOI: 10.1021/acs.biomac.9b00160

[25] Opálková Šišková A, Bučková M, Kronekova Z, Kleinová A, Nagy Š, Rydz J, et al. The drug-loaded electrospun poly(ε-caprolactone) mats for therapeutic application. Nanomaterials. 2021;**11**:922. DOI: 10.3390/nano11040922

[26] Contreras-Martínez J, García-Payo C, Khayet M. Electrospun nanostructured membrane engineering using reverse osmosis recycled modules: Membrane distillation application. Nanomaterials. 2021;**11**:1601. DOI: 10.3390/nano11061601

[27] Xu L, Liu X, Jia J, Wu H, Xie J, Jia Y. Electrospun nanofiber membranes

from 1,8-naphthimide-based polymer/ poly(vinyl alcohol) for pH fluorescence sensing. Molecules. 2022;**27**:520. DOI: 10.3390/molecules27020520

[28] Liu L, Xu T, Gui X, Gao S, Sun L, Lin Q, et al. Electrospun silsequioxane-grafted PVDF hybrid membranes for high-performance rechargeable lithium batteries. Composites Part B: Engineering. 2021;**215**:108849. DOI: 10.1016/j.compositesb.2021.108849

[29] Bayan MAH, Taromi FA, Lanzi M, Pierini F. Enhanced efficiency in hollow core electrospun nanofiber-based organic solar cells. Scientific Reports. 2021;**11**:21144. DOI: 10.1038/s41598-021-00580-4

[30] Rydz J, Opálková Šišková A, Zawidlak-Węgryńska B, Duale K. High-performance polymer applications for renewable energy. In: Devasahayam S, Mustansar C, editors. Nano Tools and Devices for Enhanced Renewable Energy. 1st ed. Amsterdam: Elsevier Science Publishing Co Inc.; 2021. pp. 3-26. DOI: 10.1016/B978-0-12-821709-2.00001-3

[31] Dankeaw A, Gualandris F, Silva RH, Scipioni R, Hansen KK, Ksapabutr B, et al. Highly porous Ce-W-TiO2 free-standing electrospun catalytic membranes for efficient de-NOx via ammonia selective catalytic reduction. Environmental Science. Nano. 2019;**6**:94-104. DOI: 10.1039/C8EN01046C

[32] Kumarage S, Munaweera I, Kottegoda N. A comprehensive review on electrospun nanohybrid membranes for wastewater treatment. Beilstein Journal of Nanotechnology. 2022;**13**:137-159. DOI: 10.3762/bjnano.13.10

[33] He ZY, Pan ZJ. Biobased polymer SF/PHBV composite nanofiber membranes as filtration and protection materials. The Journal of the Textile Institute. 2021. DOI: 10.1080/00405000.2021.2020419

[34] Žagar E, Češarek U, Drinčić A, Sitar S, Shlyapnikov IM, Pahovnik D. Quantitative determination of PA6 and / or PA66 content in polyamide-containing wastes. ACS Sustainable Chemistry & Engineering. 2020;**8**:11818-11826. DOI: 10.1021/acssuschemeng.0c04190

[35] Matulevicius J, Kliuciniskas L, Martuzevicius D, Krugly E, Tichonovas M, Baltrusaitis J. Design and characterization of electrospun polyamide nanofiber media for air filtration applications. Journal of Nanomaterials. 2014;**2014**:859656. DOI: 10.1155/2014/859656

[36] Sambaer W, Zatloukal M, Kimmer D. 3D air filtration modeling for nanofiber based filters in the ultrafine particle size range. Chemical Engineering Science. 2012;**82**:299-311. DOI: 10.1063/1.3604492

[37] Guo Y, Guo Y, He W, Zhao Y, Shen R, Liu J, et al. PET/TPU nanofiber composite filters with high interfacial adhesion strength based on one-step co-electrospinning. Powder Technology. 2021;**387**:136-145. DOI: 10.1016/j.powtec.2021.04.020

[38] Baselga-Lahoz M, Yus C, Arruebo M, Sebastián V, Irusta S, Jimenéz S. Submicronic filtering media based on electrospun recycled PET nanofibers: Development, characterization, and method to manufacture surgical masks. Nanomaterials. 2022;**12**:925. DOI: 10.3390/nano12060925

[39] Tang H, Han D, Zhang J. Electrospinning fabrication of polystyrene-silica hybrid fibrous membrane for high-efficiency air filtration. Nano Express. 2021;**2**:020017. DOI: 10.1088/2632-959X/abfe3d

[40] Bortolassi ACC, Nagarajan S, de Araújo LB, Guerra VG, Lopes Aguiar M, Huon V, et al. Efficient nanoparticled removal and bactericidal action of electrospun nanofibers membranes for air filtration. Materials Science and Engineering: C. 2019;**102**:718-729. DOI: 10.1016/j.msec.2019.04.094

[41] Molnár K, Mészáros L. The role of electrospun nanofibers in the fight against the COVID-19. Express Polymer Letters. 2020;**14**:605. DOI: 10.3144/expresspolymlett.2020.49

[42] Rajak A, Hapidin DA, Iskandar F, Munir M, Khairurrijal K. Controlled morphology of electrospun nanofibers from waste expanded polystyrene for aerosol filtration. Nanotechnology. 2019;**30**:425602. DOI: 10.1088/1361-6528/ab2e3b

[43] Strain IN, Wu Q, Pourrahimi AM, Hedenqvist MS, Olsson RT, Andersson RL. Electrospinning of recycled PET to generate tough mesomorphic fibre membranes for smoke filtration. Journal of Materials Chemistry A. 2015;**2**:1632-1640. DOI: 10.1039/C4TA06191H

[44] Opálková Šišková A, Frajová J, Nosko M. Recycling of poly(ethylene terephthalate) by electrospinning to enhanced the filtration efficiency. Mater. Letter. 2020;**278**:128426. DOI: 10.1016/j.matlet.2020.128426

[45] Bonfim DPF, Cruz FGS, Bretas RES, Guerra VG, Aguiar ML. A sustainable recycling alternative: Electrospun PET-membranes for air nanofiltration. Polymers. 2021;**13**:1166. DOI: 10.3390/polym13071166

[46] Hossain T, Ali A, Islam R, Chowdhury RH. Development of Electrospun Nanofibrous Facemask from Recycled PET Bottle Using Electrospinning Technique. Bangladesh: Textile Today, An Innovation Hub; 2021. Available from: https://www.textiletoday.com.bd/development-of-electrospun-nanofibrous-facemask-from-recycled-pet-bottle-using-electrospinning-technique/. [Accessed: May 21, 2022]

[47] Bonfim DPF, Cruz FGS, Guerra VG, Aguiar ML. Development of filter media by electrospinning for air filtration of nanoparticles from PET bottles. Membranes. 2021;**11**:293. DOI: 10.3390/membranes11040293

[48] Opálková Šišková A, Mosnáčková K, Hrůza J, Frajová J, Opálek A, Bučková M, et al. Electrospun poly(ethylene terephthalate)/silk fibroin composite for filtration application. Polymers. 2021;**13**:2499. DOI: 10.3390/polym13152499

[49] Xiao J, Lv W, Song Y, Zheng Q. Graphene/nanofiber aerogels: Performance regulation towards multiple applications in dye adsorption and oil/water separation. Chemical Engineering Journal. 2018;**338**:202-210. DOI: 10.1016/j.cej.2017.12.156

[50] Ma P, Yu Y, Fu Z. Ag3PO4/CuO composites utilizing the synergistic effect of photocatalysis and Fenton-like catalysis to dispose organic pollutants. Advanced Powder Technology. 2017;**28**(11):2797-2804. DOI: 10.1016/j.apt.2017.08.004

[51] Xiao X, Sun Y, Sun W, Shen H, Zheng H, Xu Y, et al. Advanced treatment of actual textile dye wastewater by Fenton-flocculation process. The Canadian Journal of Chemical Engineering. 2016;**95**(7):1245-1252. DOI: 10.1002/cjce.22752

[52] Liu L, Liu Z, Bai H, Sun DD. Concurrent filtration and solar photocatalytic disinfection/degradation using high-performance Ag/TiO2

nanofiber membrane. Water Research. 2012;**46**(4):1101-1112. DOI: 10.1016/j.watres.2011.12.009

[53] Tanudjaja HJ, Hejase CA, Tarabara VV, Fane AG, Chew JV. Membrane-based separation for oily wastewater: A practical perspective. Water Research. 2019;**156**:347-365. DOI: 10.1016/j.watres.2019.03.021

[54] Chen Y, Shen C, Wang J, Xiao G, Luo G. Green synthesis of Ag-TiO2 supported on porous glass with enhanced photocatalytic performance for oxidative desulfurization and removal of dyes under visible light. ACS Sustainable Chemistry & Engineering. 2018;**6**(10):13276-13286. DOI: 10.1021/acssuschemeng.8b02860

[55] Luo G, Du L, Wang Y, Wang K. Recent developments in microfluidic device-based preparation, functionalization, and manipulation of nano- and micro-materials. Particualogy. 2019;**45**:1-19. DOI: 10.1016/j.partic.2018.10.001

[56] Sui J, Yan J, Liu D, Wang K, Luo G. Continuous synthesis of nanocrystals via flow chemistry technology. Small. 2020;**16**(15):1-23. DOI: 10.1002/smll.201902828

[57] Ivshina IB, Kuyukina MS, Krivoruchko AV, Elkin AA, Makarov SO, Cuningham CJ, et al. Oil spill problems and sustainable response strategies through new technologies. Environmental Science: Processes and Impacts. 2015;**17**:1201-1219. DOI: 10.1039/c5em00070j

[58] Huang J, Yan Z. Adsorption mechanism of oil by resilient graphene aerogels from oil-water emulsion. Langmuir. 2018;**34**(5):1890-1898. DOI: 10.1021/acs.langmuir.7b03866

[59] Ma W, Samal SK, Liu Z, Xiong R, Smedt SC, Bhushan B, et al. Dual pH- and ammonia-vapor-responsive electrospun nanofibrous membranes for oil-water separations. Journal of Membrane Science. 2017;**537**:128-139. DOI: 10.1016/j.memsci.2017.04.063

[60] Zhou W, Li S, Liu Y, Xu Z, Wei S, Wang G, et al. Dual Superlyophobic copper foam with good durability and recyclability for high flux, high efficiency, and continuous oil-water separation. ACS Applied Materials & Interfaces. 2018;**10**(11):9841-9848. DOI: 10.1021/acsami.7b19853

[61] Liu W, Cui M, Shen Y, Zhu G, Luo L, Li M, et al. Waste cigarette filter as nanofibrous membranes for on-demand immiscible oil/water mixtures and emulsions separation. Journal of Colloid and Interface Science. 2019;**549**:114-122. DOI: 10.1016/j.jcis.2019.04.057

[62] Sow PK, Ishita, Singhal R. Sustainable approach to recycle waste polystyrene to high-value submicron fibers using solution blow spinning and application towards oil-water separation. Journal of Environmental Chemical Engineering. 2020;**8**(2):102786. DOI: 10.1016/j.jece.2018.11.031

[63] Zander NE, Gillan M, Sweetser S. Recycled PET nanofibers for water filtration applications. Materials. 2016;**9**:1-10. DOI: 10.3390/ma9040247

[64] Doan HN, Phong Vo P, Hayashi K, Kinashi K, Sakai W, Tsutsumi N. Recycled PET as a PDMS-functionalized electrospun fibrous membrane for oil water separation. Journal of Environmental Chemical Engineering. 2020;**8**:103921. DOI: 10.1016/j.jece.2020.103921

[65] Wang Y, Lai C, Hu H, Liu Y, Fei B, Xin JH. Temperature-responsive nanofibers for controllable oil/ water separation. RSC Advances. 2015;**5**(63):51078-51085. DOI: 10.1039/ c5ra08851h

[66] Li JJ, Zhou YN, Luo ZH. Smart Fiber membrane for pH-induced oil/water separation. ACS Applied Materials & Interfaces. 2015;**7**(35):19643-19650. DOI: 10.1021/acsami.5b04146

[67] Wang Y, Lai C, Wang X, Liu Y, Hu H, Guo Y, et al. Beads-on-string structured nanofibers for smart and reversible oil/water separation with outstanding antifouling property. ACS Applied Materials & Interfaces. 2016;**8**(38):25612-25620. DOI: 10.1021/ acsami.6b08747

[68] Xu L, Liu N, Cao Y, Lu F, Chen Y, Zhang X, et al. Mercury ion responsive wettability and oil/water separation. ACS Applied Materials & Interfaces. 2014;**6**(16):13324-13329. DOI: 10.1021/ am5038214

[69] Kwon G, Kota AK, Li Y, Sohani A, Marby JM, Tuteja A. On-demand separation of oil-water mixtures. Advanced Materials. 2012;**24**(27):3666-3671. DOI: 10.1002/ adma.201201364

[70] Li J, Li D, Yang Y, Li J, Zha F, Lei Z. A prewetting induced underwater superoleophobic or underoil (super) hydrophobic waste potato residue-coated mesh for selective efficient oil/ water separation. Green Chemistry. 2016;**18**(2):541-549. DOI: 10.1039/ c5gc01818h

[71] Du X, You S, Wang X, Wang Q, Lu J. Switchable and simultaneous oil/ water separation induced by prewetting with a superamphiphilic self-cleaning mesh. Chemical Engineering Journal. 2017;**313**:398-403. DOI: 10.1016/j. cej.2016.12.092

[72] Baggio A, Doan HN, Vo PP, Kinashi K, Sakai W, Tsutsumi N, et al. Chitosan-functionalized recycled polyethylene terephthalate Nanofibrous membrane for sustainable on-demand oil-water separation. Global Challenges. 2021;**5**(4):2000107. DOI: 10.1002/ gch2.202000107

[73] Xu R, An XC, Dar R, Xu K, Xing YL, Hu YX. Application of electrospun nanofibrous amphiphobic membrane using low-cost poly(ethylene terephthahalate) for robust membrane distillation. Journal of Water Process Engineering. 2020;**36**:101351. DOI: 10.1016/j.jwpe.2020.101351

[74] Arahman N, Fahrina A, Amalia S, Sunarya R, Mulyati S. Effect of PVP on the characteristic of modified membranes made from waste PET bottles for humic acid removal. F1000Research. 2017;**668**(6):1-14. DOI: 10.12688/ f1000research.11501.1

[75] Pulido BA, Habboub OS, Aristizabal SL, Szekely G, Nunes SP. Recycled poly(ethylene terephthalate) for high temperature solvent resistant membranes. ACS Applied Polymer Materials. 2019;**1**(9):2379-2387. DOI: 10.1021/acsapm.9b00493

[76] Kusumocahyo SP, Ambani SK, Kusumadewi S, Sutanto H, Widiputri DI, Kartawiria IS. Utilization of used polyethylene terephthalate (PET) bottles for the development of ultrafiltration membrane. Journal of Environmental Chemical Engineering. 2020;**8**(6):104381. DOI: 10.1016/j. jece.2020.104381

[77] Kusumocahyo SP, Ambani SK, Marceline S. Improved permeate flux and rejection of ultrafiltration membranes

prepared from polyethylene terephthalate (PET) bottle waste. Sustainable Environment Research. 2021;**31**(19):1-11. DOI: 10.1186/s42834-021-00091-x

[78] Kiani S, Mousavi SM, Bidaki A. Preparation of polyethylene terephthalate/xantan nanofiltration membranes using recycled bottles for removal of diltiazem from aqueous solution. Journal of Cleaner Production. 2021;**314**:128082. DOI: 10.1016/j.jclepro.2021.128082

[79] Siegel JD, Ko CJ. Diltiazem-associated photodistributed hyperpigmentation. Yale Journal of Biology and Medicine. 2020;**93**(1):45-47

[80] Attia AAM, Abas KM, Nada AAA, Shouman MAH, Opálková Šišková A, Mosnáček J. Fabrication, modification, and characterization of lignin-based electrospun fibers derived from distinctive biomass sources. Polymers. 2021;**13**:2277. DOI: 10.3390/polym13142277

[81] Yasin SA, Zeebaree SYS, Zeebaree AYS, Zebari OIH, Saeed IA. The efficient removal of methylene blue dye using CuO/PET nanocomposite in aqueous solutions. Catalysts. 2021;**11**:241. DOI: 10.3390/catal11020241

[82] Datsyuk V, Trotsenko S, Peikert K, Hoeflich K, Wedel N, Allar C, et al. Polystyrene nanofibers for nonwoven porous building insulation materials. Engineering Reports. 2019;**1**:e12037. DOI: 10.1002/eng2.12037

Chapter 5

# Application of Nanocellulose Biocomposites in Acceleration of Diabetic Wound Healing: Recent Advances and New Perspectives

*Rebika Baruah and Archana Moni Das*

## Abstract

Diabetes mellitus (DM) is a chronic health problem that increases the risk of infection and delays wound healing due to impairment of metabolic activity. Diabetic foot ulcers (DFUs), a chronic wound increases the risk of mortality. Finding the most appropriate wound dressings has been intensified with the increasing population and prevalence of chronic wounds. Nanofibers coated wound dressings have attracted more attention as innovative and biocompatible materials. Nanocellulose (NC) has been widely used as a reinforcing material to improve nanofibers' mechanical and thermal properties. NC is biodegradable and derived from renewable sources and produced bionanocomposites with improved performance.

**Keywords:** Diabetes mellitus, diabetic foot ulcers, wound dressings, nanocellulose, bionanocomposites

## 1. Introduction

Diabetes mellitus (DM) is a complex chronic metabolic and endocrine disorder that is termed a silent killer. DM is distinguished by a higher level of glucose in the blood which is termed hyperglycemia [1]. DM is classified into two classes and they are TYPE I DM and TYPE II DM. According to International Diabetes Federation (IDF), the estimated number of diabetic adult patients (20–75) worldwide is 463 million, and it will be 578.4 million by 2030 and 700.2 million by 2045 as per expectations. Based on the 2019 prevalence data of IDF, the number of death and complications that arise due to DM worldwide is 4.2 million [2].

Diabetic foot ulcers (DFU) are a chronic infection that arises due to DM. DM decreases the rate of wound healing due to impairment of metabolic activity. Annually, 9.1–26.1 million people were infected by DFU and many cases were selected for amputation as an ultimate treatment. DFU amputees have a 3 years survival due to the risk of infection and unsolved artery injury and 50–70% of patients have a recurrence in 5 years. Therefore, in the prevention of DFU of DM patients should maintain a lot of precautions in their lifestyles, and time consuming and high level of treatment by a multidisciplinary group of specialists is required [3].

The need for highly efficient approachable wound dressing materials is very essential in the treatment of DFU with the increasing of populations. Nanofibrous composites are attractive wound dressing materials due to their similarity with the extracellular matrix of the skin. They form an effective layer over a wound that stimulates the tissue to fight against infections causing pathogens and accelerate the wound healing [4]. Nanocellulose (NC) is used as a reinforcing material to improve the mechanical and thermal properties of nanofibrous wound dressing materials. Due to the small size, high surface area, high mechanical strength, biodegradability, and renewability of nanocellulose, it facilitates the production of cost-effective and eco-friendly wound dressing materials [5].

Several bionanocomposites of nanocellulose along with other components are applied in diabetic wound therapy in recent years [6]. Bionanocomposites are nanohybrids composed of biobased materials and inorganic components. In bionanocomposites one component should be in nano dimensions, e.g., nanopolymer, inorganic NPs, etc. [7]. Bionanocomposites cover an intensive area of research due to their biocompatibility, nontoxicity, biodegradability, renewability, and cost-effective nature. Among them, cellulose-based nanocomposites have attracted researchers in the biomedical field due to the abundance of cellulose in nature and their biocompatible nature [8]. Keeping this in mind, many researchers developed nanocellulose reinforcing biocompatible patches, bionanocomposites hydrogel for diabetic wound healings. The objectives of this chapter are to study the mechanism of normal wound healing and diabetic wound healing, different methods of synthesis of nanocellulose based bionanocomposites, and application of different bionanocomposites of cellulose in diabetic wound healing. This chapter also discussed the future perspectives of nanocellulose based wound healing dressings in DFU treatment.

## 2. Application of naocellulose-based nanocomposites in diabetic wound healing

### 2.1 What is a wound?

Skin is the largest organ of the human body, it not only protects the human body from different pathogens and the external environment but also has a great impact on different metabolic activities of the human body. Any kind of damage to the skin or other parts of the human body is termed a wound. The wound includes cuts, scrapes, scratches, and punctured skin [9]. Although the accident is the main cause of the wound, it also happens due to surgery, sutures, and stitches. Wounds happen on the skin due to exposure to the external environment and its elastic and soft nature is susceptible to generating defects. Wounds are classified into two types- acute and chronic wounds [10]. Acute wounds happen due to bites, cuts, minor burns, and surgery. These types of wounds heal within a predictable period and follow the four stages of the healing process; they are hemostasis, inflammation, proliferation, and maturation. Chronic wounds do not follow the healing process and disturbance of the physiology of the wound happens due to various endogenous mechanisms. Chronic wounds damage the integrity of the tissue and increase the healing periods. Diabetic foot ulcers, pressure sores, etc., are examples of chronic wounds. Chronic wounds occur due to aging, malnutrition, immunosuppression diseases like AIDS [11]. Pathogens infected the wound from beached skin and infection creates pain, discoloration of the wounded area, edema, puss, smell, tenderness, etc. Microbes produce biofilm in the wound and make difficulties in the healing process. With the help of biofilm, bacteria can transfer the antibiotic-resistant gene to other bacterial species. Biofilm makes a barrier between antibiotics and bacteria and makes it tough to heal the infected wound [12]. Hence, it is very necessary to develop a new and novel wound dressing material that can heal wounds infected by multidrug-resistant bacteria.

### 2.2 Diabetic wound

DM causes severe damage to multi-organ of DM patients such as cardiovascular disease, chronic renal failure, and diabetic skin wounds [13]. Lower extremities of diabetic patients infected by multidrug-resistant pathogens and the phenomenon are called diabetic foot ulcer (DFU). Annually highest numbers of DFU patients admit to hospital and require a large amount of management cost among other severe diseases. The infected wound of DFU damages the defensive sensation associated with peripheral artery disease and also develops cell necrosis that leads to amputation [8]. 80% of lower limb amputation is the result of DFU annually [14]. According to Control Cardiovascular Risk in Diabetes (ACCORD) DM patients can reduce the risk of amputation of the lower limb by maintaining the standard glycemic control (HbA1c 53–63 mmol/mol [7.0–7.9%]) or taking antidiabetic therapy (HbA1c $<$ 42 mmol/mol [$<$6%]) [10]. DFU developed due to many factors such as loss of glycemic control, peripheral vascular disease, peripheral neuropathy, and immunosuppression [11]. Neuropathy is the main contributor to DFU. Accumulation of sorbitol and fructose in a hyperglycemic state decreases the production of nerve cell myoinositol, which interrupts normal neuron conduction [12]. This phenomenon creates an imbalance between flexion and extension of the foot and produces foot deformities. Hyperglycemia also causes vascular diseases, such as endothelial cell dysfunction and an increase in thromboxane A2 [11]. All these are the key factors for DFU and the untreatable condition of infected wound leads to amputation. Nanocomposites consisting of polymer, polysaccharides, and inorganic nanoparticles exhibit better antipathogenic properties in lower dosages than conventional drugs and easily overcome the complications of multidrug-resistant pathogens [3]. Here, this chapter would focus on the application of nanocellulose based composites in the treatment of DFU. Along with this, a brief description and mechanism of normal wound healing and diabetic wound healing was tried to provide for better understanding.

### 2.3 Wound healing

The healing of a wound proceeds through a complex pathway with dynamic interactions of different cell types, extracellular matrix (ECM), cytokines, and growth factors. Fundamentally, wound healing consists of four steps, hemostasis, inflammation, cell movement, and proliferation, followed by wound compression and further remodeling [14]. Hemostasis involves the clotting process through the coagulation cascade that leads to the cessation of bleeding. Platelets are the first subset of cells that enter the injury site and release several growth factors such as platelet-derived growth factor (PDGF), transforming growth factor-beta, endothelial growth factor (EGF), and fibroblast growth factor (FGF), which support the inflammation process [15]. Hemostasis is followed by inflammation through vascular delivery of inflammatory agents and migration of cells into the injury site. Inflammatory mediators such as prostaglandins, histamine, and leukotrienes by mast cells stimulate angiogenesis and permeability to allow cells and molecules from the bloodstream to enter the wound site [16]. White blood cells such as neutrophils, monocytes, and lymphocytes invade the injury site. Neutrophils prevent microbial infections and macrophages, secret TGF, vascular endothelial growth factor (VEGF), and FGF to stimulate angiogenesis. Neutrophils also produce tumor necrosis factor-alpha (TNF) that breakdown necrotic tissue and facilitate the proliferation of fibroblasts for deposition of collagen for tissue granulation [17]. The dermal wound is followed by wound contraction after 2 weeks. Fibroblasts convert to myofibroblasts phenotype during tissue granulation, with enhanced alpha-smooth muscle actin cytoskeleton that plays a vital role in wound closure. During the re-epithelialization of tissue, keratinocytes migrate, differentiate, and proliferate to generate a stratified epidermis along the superficial area of injury to provide cover for newly formed tissue and new tissue covers wound bed [18]. Remodeling is the last phase of wound healing that lasts for 6–24 months. Remodeling involves the generation of granulation tissue accompanied by the replacement of the ECM with type I collagen (substituting collagen III) mediated via PDGF and TGF [19].

## 2.4 Diabetic wound healing

Inflammatory macrophages stay at the site of injury in the case of diabetic wounds for a longer period compared to normal wound healing. Increased ratio of pro-inflammatory cytokines, such as TNF and interleukin 6 (IL-6) are produced by inflammatory macrophages. Macrophages also elaborate ROS that causes persistent inflammation and lead to stimulation of proliferative factors for successful wound healing. In the case of diabetic wounds, inefficient efferocytosis (phagocytosis of apoptotic cells) by macrophages perturbed cytokine cascade and causes a higher burden of apoptotic cells. Pro-inflammatory cytokines (IL-1 and TNF) and matrix metalloproteinase-9 (MMP-9) increase in ratio with decreased anti-inflammatory signals (CD206, IGF-1, TGF, and IL-10) and lead to abnormal apoptosis of fibroblasts and keratinocytes, and decreased angiogenesis [20]. Fibroblasts do not properly convert to myofibroblasts in diabetic wound healing, which reduces mechanical tension of ECM, and results in poor wound closure due to lack of SMA [21]. Again, a non-equilibrium balance between MMPs degrades the disorganized collagen in normal wound healing and tissue inhibitor of metalloproteinases (TIMPs), as a result, degradation and deposition of abnormal ECM occur. High levels of pro-inflammatory cytokines and pro-fibrotic cytokines reduced the level of expression of TIMPs and higher expression of MMPs. Levels of MMPs are raised 60 times more in chronic wounds than for acute wound healing [22]. In chronic wounds, degradation of ECM, growth factors, and collagen deposition increases due to an increase in protease activity in tissue reconstruction [22]. All these factors play a crucial role in wound healing. Along with these factors, a dysregulated molecular and cellular wound microenvironment is not conducive to normal healing responses and culminates in impaired healing of diabetic ulcers.

## 2.5 Nanocellulose

Bio-renewable materials are the most demandable factors in the construction of a sustainable planet. Biopolymer is the promising substitute for the petroleum-based product in the environment safe concern. Natural feedstocks such as agricultural waste, food waste, wood, etc., can be converted to value-added products. Plants cells are composed of cellulose (40–50 wt%), hemicellulose (20–40 wt%), and lignin (20–30%), and these lignocellulosic materials are the most abundant bio-resources on the earth for sustainable and renewable products [23]. Cellulose is the most abundant natural polymer on earth. Chemically, cellulose is composed of a linear chain of homo-polysaccharide glucose units linked by a $\beta$-1,4-glycosidic bond. Nano form of cellulose is termed nanocellulose. Cellulose can be converted to nanocellulose by breaking the intra- and intermolecular hydrogen bond between the polymeric chains of the cellulose [24]. Nanocellulose can be sourced from plants, microorganisms, and aquatic animals (tunicates), rich in cellulose. Different plants such as banana leaf, corn cob, cotton, ramie, rice husk, wood, sugarcane bagasse, sisal leaves, wheat straw, wood, and coconut husks are rich sources of nanocellulose. Nanocellulose can be also extracted from coffee grounds, ginger, durian rind waste, lemon seeds, okara, pea hull, *Phragmites australis*, *Hevea brasiliensis* (Rubberwood), and tea stalk. Nanocellulose extracted from bacteria is known as bacterial nanocellulose (BNC) or bacterial cellulose (BC). *Acetobacterxylinum* (*Gluconacetobacter xylinus*) is the most efficient bacteria used for nanocellulose production. Other bacterial species *Acetobacter, Achromobacter, Agrobacterium, Acanthamoeba, Alcaligenes, Rhizobium, Pseudomonas, Sarcina*, and algal species *Cladophora, Rhizoclonium, Microdiction,* and *Chaetomorpha*, are also the potential source of nanocellulose. BC is purer than nanocellulose derived from other sources. But the molecular structure of BC and plant-derived nanocellulose are similar [5].

#### *2.5.1 Classification of nanocellulose*

Nanocellulose is classified into three main classes based on its morphological structure and sources. They are cellulose nanocrystals (CNCs/NCC), cellulose nanofibers (CNFs/NFC), and bacterial nanocellulose (BNC/BC).

##### *2.5.1.1 Cellulose nanocrystals (CNCs/NCC)*

CNCs are extracted from cellulose-rich sources. Acid treatment is mainly utilized to convert cellulose to nanocellulose. They are highly crystalline with a needle-like structure. Their dimensions are 4–20 nm in width and 100–500 nm in length.

##### *2.5.1.2 Cellulose nanofibers (CNFs/NFC)*

Chemical treatment and mechanical disintegration are utilized to extract cellulose. nanofibers (CNFs). They are both crystalline and amorphous with 1 μm in length.

##### *2.5.1.3 Bacterial nanocellulose (BNC/BC)*

Nanocellulose originates from microorganisms defined as BNC. They are the purest form of nanocellulose. Their diameter range is 20–100 nm and several micrometers in length [5].

### 2.6 Nanocellulose based composites

Nanocellulose is the promising reinforcing agent and filler in the formation of bionanocomposites. Due to its unique tensile strength, thermal behavior, and ease of surface modifications, it becomes a potential constituent in the development of multipurpose nanocomposites. Bionanocomposites are an efficient applicant in various fields due to their biodegradability, biocompatibility, and cost-effective nature, and they also possess a high surface-to-volume ratio and nanometric size effect. Nanocellulose seeks the attention of researchers due to its abundances, renewability, environmentally friendly, cost-effective, and improved mechanical nature. Nanocomposites are classified into two categories; they are natural and synthetic nanocomposites based on the constituents [25].

## 2.7 Biomedical application of nano-cellulose based bionanocomposites

Nanocellulose is a potential applicant in biomedical fields due to its biocompatibility, biodegradability, high surface area, low density, high mechanical, thermal, and optical properties [26]. Nanocellulose can be efficiently applied in drug delivery, tissue engineering, wound healing, and as antimicrobial agents [27]. Nanocellulose-based nanocomposites show excellent activity in the delivery of loaded drugs in the tropical site. This factor helps in the development of nanocellulose-based wound dressing in the treatment of normal as well as chronic wound (DFU) [28].

### *2.7.1 Application of nanocellulose-based nanocomposites in wound healing*

The bacterially infected wound is very hard to repair due to the biofilm formation by bacteria, drug resistance, and high recurrence properties of bacteria [29]. Among bacteria, Staphylococcus species were responsible for around 70% of bone-related infections. The formation of biofilm with high adherence and the inefficient drug release system delay wound healing [30]. Nanocomposites are sustainable and efficient antimicrobial drug delivery systems that overcome the obstacles from implant-associated bacterial infections [31]. E.g., Starch/Triphala Churna (TC) nanocomposites exhibited high drug loading efficacy and prolonged TC release for improved acetylcholinesterase inhibition. These nanocomposites also possess excellent antimicrobial activity with antioxidant activity against multidrug resistance biofilm-forming human bacterial pathogens [32].

Cellulose materials are the most appealing candidate in the development of eco-friendly and durable materials [33]. Nanocomposites of nanocellulose have clinical potential applications in tissue engineering. Various nano-scaffold of nanocellulose mimic the nature of the tissue and substitute degenerated tissues. CNC/folic acid delivered targeted chemotherapeutic agents to folate receptor-positive 886 cancer cells [34]. CNC/hydroquinone inhibited the production of melanin, was designed for treating hyperpigmentation, a disorder occurring during pregnancy and sun exposure [35]. Various cellulose-based nanocarriers, such as bacterial cellulose, cellulose acetate, microcrystalline cellulose, carboxymethyl cellulose, cellulose nanocrystals, cellulose nanofibrils, etc., in drug delivery systems for cancer treatment, were used [36]. Nanocellulose-based wound dressings were combined with biosensors, e.g., for human neutrophil elastase was used in nanocellulose based nanocomposite in chronic wound healing [37].

Advantages of the biodegradable polymer in wound dressing overcome the obstacles of not degradable polymer containing wound dressing material in the case of removal of wound dressings after healing. The human body is lacking the cellulose enzyme of bacteria that relate to the biodegradability of the cellulose polymer. Modified or derivative forms of cellulose-containing wound dressing materials solve this problem. E.g., a glucose polymer of dextran was used to improve the biodegradability of BC and cell proliferation in wound site [38] and Chloramphenicol/2,3-dialdehyde cellulose hydrogel showed higher biodegradability, drug delivery ability, and efficient antibacterial actions against *E. coli*, *S. aureus*, and *Streptococcus pneumoniae* [39]. Gram-positive (*S. aureus* and *Enterococcus faecalis*) and gram-negative bacteria (*P. aeruginosa*) are significant in single or multi-bacterial wound infections [40]. Sufficient moisture, oxygen, temperature, growth factors, and bioactive materials play a vital role in the fast healing of the wound [41]. BC/ZnO nanocomposites more effectively repair burn wounds (77%) than sulfadiazine (66%) within fifteen days [42, 43]. Among nanocellulose, NCC and NFC are specific in wound healing due to their high degree of functionality and biocompatibility. Sodium periodate oxidized NFC was applied in the integration of collagen polymer. This nanocomposite exhibited high water absorption (4000%), porosity (95%), and density of collagen/NFC (0.03 g/cm$^3$) [44]. In another study, periodate and TEMPO were used to oxidize and carboxymethylated NC for use as a bioink material with growth inhibition ability in the case of *P. aeruginosa* PAO1 [45]. Mechanical and cytotoxic properties of nanocomposites of NCCs showed significant water absorbance with biocompatibility effect on adipose-derived stem cells (ASCs) and L929 cell line after 7 days [46] as compared to chitosan polymer. CNC/chitosan/polyethylene oxide composites showed increased tensile strength, tensile modulus, and $O_2$/$CO_2$ transmission and had no cytotoxic impact on adipose-derived stem cells. Hence, the nanocomposites were considered promising applicants for wound healing [47]. CNC composed wound dressings also promote coagulation processes in bleeding wounds [48]. E.g., biodegradable nanocomposites film and sponges of oxidized CNC/alginate exhibited excellent hemostatic efficiency without eliciting an inflammatory reaction. Carboxyl functionalization on the CNC surface was responsible for the hemostatic effect of the nanocomposites. Due to the hemostatic effect, blood plasma was significantly absorbed in the material science and it became more effective in promoting platelet aggregation and in inducing erythrocytes to accelerate blood clotting [49]. Curcumin and gelatin microspheres (GMs) were loaded with scaffolds of porous collagen (Coll)-CNC composite. These scaffolds successfully released curcumin and exhibited potential antibacterial activity in vitro. They also effectively repair the full-thickness burn infections through accelerating dermis regeneration and preventing local inflammation [50]. CNC/chitosan/Ag NPs/curcumin nanocomposites accelerated the complete healing of excision wounds in albino rats without skin irritation [51]. Gentamycin sulfate (GS)/collagen (Coll)/CNC/gelatin microspheres (GMs) scaffolds and CNC/poly(N-isopropyl acrylamide) (PNIPAAm)/metronidazole hydrogel are promising wound dressing with antibacterial activity [52]. CNC/alginate hydrogel consisting of a double membrane system of cationic CNC (CCNC) and anionic alginate was applied to design hydrogel to incorporate different drugs in each hydrogel membrane: ceftazidime hydrate (CH) antibiotics in the external membrane (pure alginate) and EGF in the internal membrane (CCNC and alginate). This hydrogel was suitable for wound dressing with rapid drug release and oral drug delivery applications [53]. Electrospinned polyethylene imine carboxymethyl chitosan/pDNA-angiogenin (ANG) nanoparticles/curcumin (Cur)/poly(D,L-lactic-co-glycolic acid) (PLGA)/CNCs nanodispersion was developed to deliver a novel drug and plasmid DNA (pDNA). The composites exhibited potential antioxidant, antitumor, anti-inflammatory, and antibacterial activity and repaired full-thickness burn wounds by stimulating blood vessel formation and sustained release of drugs [54].

### 2.8 Application of nanocellulose based nanocomposites in diabetic foot ulcer (DFU)

Special wound dressing is required to repair DFU which has prolonged drug release properties, maintains optimum moist conditions, and protects wound from infection. Polymers and polysaccharides based nanostructured wound dressing materials are ideal for DFU treatment due to their notable biocompatibility, high drug loading capacity, and accessible tailoring properties [55]. Electrospinned berberine/cellulose acetate (CA)/gelatin mat was developed as a potential wound dressing for diabetic foot ulcers [56]. CNC extracted from *Syzygium cumini* was applied to CNC/Ag NPs nanocomposites in two different forms. In ointment form, it repairs wound in diabetic mice by promoting increased collagen deposition, angiogenesis and enhanced the formation of neo-epithelialization. The wound healing capability was better in ointment form than in strip form. Again, CNC derived from *Syzygium cumini* leaves or bamboo was utilized in the synthesis of CNC/Ag NPs and applied in the repairing of acute and diabetic wounds [57]. PLGA/CNC nanofibres loaded with neurotensin (inflammatory modulator) were applied as potential wound dressings for accelerated diabetic wound healing [58]. CNC/curcumin nanocomposite films were applied to diabetic wounds showed stable curcumin release, which plateaued after 36 h [59]. Antimicrobial CNC/Ag nanocomposite wound dressings with excellent water absorption capacities were applied in diabetic or full-thickness wounds. They reduced pro-inflammatory cytokine levels and completely recovered the wound by elevating collagen and improving re-epithelization and vasculogenesis [60]. Biosensor dressing of CNF film loaded with Venlafaxine (an analgesic drug used to treat neuropathic pain) and tetracycline (antibiotic) was applied simultaneous delivery of two drugs to diabetic foot ulcers. The CNF film served as an appropriate carrier for the co-delivery of the two drugs [61].

## 3. Conclusion and future perspectives

This chapter provides a state-of-the-art review of the application of nanocomposites of nanocellulose in the biomedical field, especially in diabetic wound healing as sustainable wound dressings. Nanocellulose has tremendous applications in various fields such as biomedical, environmental remediation, food technology, electronics, etc. Due to the natural availability, biodegradability, biocompatibility, and tunable nature nanocellulose is an effective drug carrier in target-specific delivery of the drug. Incorporation of other NPs or impregnation of nanocellulose in the nanocomposites enhances the drug loading capacity of nanocellulose. The dual drug delivery ability of nanocellulose is more efficient than the single drug delivery. This property is useful for the delivery of various combinations of drugs in the treatment of various chronic diseases. The potential application of nanocellulose-based composites in the biomedical field is studied, but the in-depth toxicological properties of nanocellulose based composites have not been evaluated yet. This study will be helpful for the more advanced clinical application of nanocellulose based composites.

Diabetic wound healing is a complex and prolonged process due to its pathophysiology, resulting in the impaired function of different cells and unbalanced levels of key biochemical healing mediators. Advanced nanowound dressings are applied to overcome the challenges of diabetic wound healing. Nanocomposites are synthesized for this application by combinations of biomolecules (such as growth factors, genes, proteins/peptides, stem cells/exosomes) and non-bioactive substances (metal ions, oxygen, and nitric oxide), as well as nanotechnology (e.g., PTT, LBL self-assembly technique and 3D printing). The etiopathogenesis of diabetic foot ulcers is too complex and challenging. Wound dressings should consist of more than two materials. Therefore, potential constituents should be applied that have antimicrobial as well as biocompatible nature. Nanocellulose based nanocomposites are tried to apply in wound healing of diabetic patients that have fulfilled all the criteria of wound dressings of DFU. Future directions will try to develop new nanocellulose based wound dressings with multiple roles (including improving hypoxia, enhancing angiogenesis, reducing oxidative stress, and preventing infection). These types of nano dressings will accelerate diabetic wound healing in all stages and maintain a balanced environment during wound healing. Nanocellulose based nanocomposites will be believed to be nano wound dressings that will reduce all the complications in the treatment of DFU and fully cure diabetic wounds in a short period without any side effects.

## Acknowledgements

The authors sincerely acknowledged the Director of CSIR-North East Institute of Science and Technology, Jorhat, Assam for his permission to perform our research with excellent facilities and precious guidance. R. B. acknowledged the Department of Science and Technology, New Delhi, for the Inspire Fellowship (GAP-0754).

## Conflict of interest

The authors declare no conflict of interest.

## Author details

Rebika Baruah[1,2]* and Archana Moni Das[1,2]

1 Natural Products Chemistry Group, Chemical Science and Technology Division, CSIR-North East Institute of Science and Technology, Jorhat, Assam, India

2 Academy of Scientific and Innovative Research (AcSIR), Ghaziabad, India

*Address all correspondence to: baruahrebika9@gmail.com

## References

[1] Choudhury H, Pandey M, Lim YQ, Low CY, Lee CT, Marilyn TCL, et al. Silver nanoparticles: Advanced and promising technology in diabetic wound therapy. Materials Science and Engineering: C. 2020;**112**:110925. DOI: 10.1016/j.msec.2020.110925

[2] Ezhilarasu H, Vishalli D, Dheen ST, Bay BH, Srinivasan DK. Nanoparticle-based therapeutic approach for diabetic wound healing. Nanomaterials. 2020; **10**:1234

[3] Wu SC, Driver VR, Wrobel JS, Armstrong DG. Foot ulcers in the diabetic patient, prevention and treatment. Vascular Health and Risk Management. 2007;**3**:65-76

[4] Chen JP, Chang GY, Chen JK. Electrospun collagen/chitosan nanofibrous membrane as wound dressing. Colloids and Surfaces A: Physicochemical and Engineering Aspects. 2008;**314**:183-188

[5] Patil TV, Patel DK, Dutta SD, Ganguly K, Santra TS, Lim KT. Nanocellulose, a versatile platform: From the delivery of active molecules to tissue engineering applications. Bioactive Materials. 2022;**9**:566-589

[6] Habibi Y, Dufresne A. Highly filled bionanocomposites from functionalized polysaccharide nanocrystals. Biomacromolecules. 2008;**9**:1974-1980. DOI: 10.1021/bm8001717

[7] Habibi Y, Goffin AL, Schiltz N, Duquesne E, Dubois P, Dufresne A. Bionanocomposites based on poly (ε-caprolactone)-grafted cellulose nanocrystals by ring-opening polymerization. Journal of Materials Chemistry. 2008;**18**:5002-5010. DOI: 10.1039/b809212e

[8] Shen R, Xue S, Xu Y, Liu Q, Feng Z, Ren H, et al. Research progress and development demand of nanocellulose reinforced polymer composites. Polymers. 2020;**12**:2113. DOI: 10.3390/polym12092113

[9] Moon RJ, Martini A, Nairn J, Simonsen J, Youngblood J. Cellulose nanomaterials review: Structure, properties and nanocomposites. Chemical Society Reviews. 2021;**40**:3941. DOI: 10.1039/c0cs00108b

[10] Bacakova L, Pajorova J, Bacakova M, Skogberg A, Kallio P, Kolarova K, et al. Versatile application of nanocellulose: From industry to skin tissue engineering and wound healing. Nanomaterials. 2019;**9**:164. DOI: 10.3390/nano9020164

[11] Shefa AA, Amirian J, Kang HJ, Bae SH, Jung HI, Choi H, et al. In vitro and in vivo evaluation of effectiveness of a novel TEMPO-oxidized cellulose nanofiber-silk fibroin scaffold in wound healing. Carbohydrate Polymers. 2017;**177**:284-296. DOI: 10.1016/j.carbpol.2017.08.130

[12] Iqbal A. Management of chronic non-healing wounds by hirudotherapy. World Journal of Plastic Surgery. 2017;**6**:9

[13] Mustoe T. Meeting: Tissue repair and ulcer/wound healing: Molecular mechanisms, therapeutic targets and future directions. In: Dermal Ulcer Healing: Advances in Understanding. 2005. p. 2005. CRID: 1570009749393111808, NII Article ID: 20001383165, CiNii Articles

[14] Han G, Ceilley R. Chronic wound healing: A review of current management and treatments. Advanced

Therapy. 2017;**34**:599-610. DOI: 10.1007/s12325-017-0478-y

[15] Sun BK, Siprashvili Z, Khavari PA. Advances in skin grafting and treatment of cutaneous wounds. Science. 2014;**346**:941-945

[16] Okonkwo UA, DiPietro LA. Diabetes and wound angiogenesis. International Journal of Molecular Sciences. 2017;**18**:1-15. DOI: 10.3390/ijms18071419

[17] Kesharwani P, Gorain B, Low SY, Tan SA, Ling ECS, Lim YK, et al. Nanotechnology based approaches for anti-diabetic drugs delivery. Diabetes Research and Clinical Practice. 2018;**136**:52-77. DOI: 10.1016/j.diabres.2017.11.018

[18] Bhattamisra SK, Siang TC, Rong CY, Annan NC, Sean EHY, Xi LW, et al. Type-3c diabetes mellitus, diabetes of exocrine pancreas—An update. Current Diabetes Reviews. 2019;**15**:382-394. DOI: 10.2174/1573399815666190115145702

[19] Hicks CW, Selvarajah S, Mathioudakis N, Sherman RL, Hines KF, Black JH, et al. Burden of infected diabetic foot ulcers on hospital admissions and costs. Annals of Vascular Surgery. 2016;**33**:149-158. DOI: 10.1016/j.avsg.2015.11.025

[20] Monteiro-Soares M, Ribeiro-Vaz I, Boyko EJ. Canagliflozin should be prescribed with caution to individuals with type 2 diabetes and high risk of amputation. Diabetologia. 2019;**62**:900-904. DOI: 10.1007/s00125-019-4861-x

[21] Pecoraro RE, Reiber GE, Burgess EM. Pathways to diabetic limb amputation: Basis for prevention. Diabetes Care. 1990;**13**:513-521. DOI: 10.2337/diacare.13.5.513

[22] Goldman MP, Clark CJ, Craven TE, Davis RP, Williams TK, Velazquez-Ramirez G, et al. Effect of intensive glycemic control on risk of lower extremity amputation. Journal of the American College of Surgeons. 2018;**227**:596-604. DOI: 10.1016/j.jamcollsurg.2018.09.021

[23] Aumiller WD, Dollahite HA. Pathogenesis and management of diabetic foot ulcers. Journal of the American Academy of Physician Assistants. 2015;**28**:28-34. DOI: 10.1097/01.JAA.0000464276.44117.b1

[24] Clayton W, Elasy TA. A review of the pathophysiology, classification, and treatment of foot ulcers in diabetic patients. Clinical Diabetes. 2009;**27**:52-58. DOI: 10.2337/diaclin.27.2.52

[25] Choudhury H, Pandey M, Yin TH, Kaur T, Jia GW, Tan SQL, et al. Rising horizon in circumventing multidrug resistance in chemotherapy with nanotechnology. Materials Science and Engineering: C. 2019;**101**:596-613 Available from: http://www.ncbi.nlm.nih.gov/pubmed/31029353.

[26] Gorain B, Tekade M, Kesharwani P, Iyer AK, Kalia K, Tekade RK. The use of nanoscaffolds and dendrimers in tissue engineering. Drug Discovery Today. 2017;**22**:652-664. DOI: 10.1016/j.drudis.2016.12.007

[27] Parayath NN, Parikh A, Amiji MM. Repolarization of tumor-associated macrophages in a genetically engineered nonsmall cell lung cancer model by intraperitoneal administration of hyaluronic acid-based nanoparticles encapsulating microRNA-125b. Nano Letters. 2018;**18**:3571-3579. DOI: 10.1021/acs.nanolett.8b00689

[28] Pastar I, Stojadinovic O, Yin NC, Ramirez H, Nusbaum AG, Sawaya A,

et al. Epithelialization in wound healing: A comprehensive review. Advanced Wound Care. 2014;**3**:445-464. DOI: 10.1089/wound.2013.0473

[29] Atala A, Irvine DJ, Moses M, Shaunak S. Wound healing versus regeneration: Role of the tissue environment in regenerative medicine. MRS Bulletin. 2010;**35**:597-606. DOI: 10.1557/mrs2010.528

[30] Evans ND, Oreffo ROC, Healy E, Thurner PJ, Man YH. Epithelial mechanobiology, skin wound healing, and the stem cell niche. Journal of the Mechanical Behavior of Biomedical Materials. 2013;**28**:97-409. DOI: 10.1016/j.jmbbm.2013.04.023

[31] Rajendran NK, Kumar SSD, Houreld NN, Abrahamse H. A review on nanoparticle based treatment for wound healing. Journal of Drug Delivery Science and Technology. 2018;**44**:421-430. DOI: 10.1016/j.jddst.2018.01.009

[32] Li Y, Wang B, Ma M, Wang B. Review of recent development on preparation, properties, and applications of cellulose-based functional materials. International Journal of Polymer Science. 2018;**2018**:1-18. DOI: 10.1155/2018/8973643

[33] Koh TJ, DiPietro LA. Inflammation and wound healing: The role of the macrophage. Expert Reviews in Molecular Medicine. 2011;**13**:e23. DOI: 10.1017/S1462399411001943

[34] Ho JK, Hantash BM. The principles of wound healing. Expert Review of Dermatology. 2013;**8**:639-658. DOI: 10.1586/17469872.2013.857161

[35] Pakyari M, Farrokhi A, Maharlooei MK, Ghahary A. Critical role of transforming growth factor beta in different phases of wound healing. Advances in Wound Care. 2013;**2**:215-224. DOI: 10.1089/wound.2012.0406

[36] Tahergorabi Z, Khazaei M. Imbalance of angiogenesis in diabetic complications: The mechanisms. International Journal of Preventive Medicine. 2012;**3**:827-838. DOI: 10.4103/2008-7802.104853

[37] Wu J, Zheng Y, Song W, Luan J, Wen X, Wu Z, et al. In situ synthesis of silver-nanoparticles/bacterial cellulose composites for slow-released antimicrobial wound dressing. Carbohydrate Polymers. 2014;**102**:762-771. DOI: 10.1016/j.carbpol.2013.10.093

[38] Tsourdi E, Barthel A, Rietzsch H, Reichel A, Bornstein SR. Current aspects in the pathophysiology and treatment of chronic wounds in diabetes mellitus. BioMed Research International. 2013;**2013**:385641. DOI: 10.1155/2013/385641

[39] Varkey M, Ding J, Tredget E. Advances in skin substitutes—Potential of tissue engineered skin for facilitating anti-fibrotic healing. Journal of Functional Biomaterials. 2015;**6**:547-563. DOI: 10.3390/jfb6030547

[40] Borys S, Hohendorff J, Frankfurter C, Kiec-Wilk B, Malecki MT. Negative pressure wound therapy use in diabetic foot syndrome—From mechanisms of action to clinical practice. European Journal of Clinical Investigation. 2019;**49**:e13067. DOI: 10.1111/eci.13067

[41] Streubel PN, Stinner DJ, Obremskey WT. Use of negative-pressure wound therapy in orthopaedic trauma. The Journal of the American Academy of Orthopaedic Surgeons. 2012;**20**:564-574. DOI: 10.5435/JAAOS-20-09-564

[42] Das S, Baker AB. Biomaterials and nanotherapeutics for enhancing

skin wound healing. Frontiers in Bioengineering and Biotechnology. 2016;**4**:82. DOI: 10.3389/fbioe.2016.00082

[43] Katepetch C, Rujiravanit R, Tamura H. Formation of nanocrystalline ZnO particles into bacterial cellulose pellicle by ultrasonic-assisted in situ synthesis. Cellulose. 2013;**20**:1275-1292

[44] Hamdan S, Pastar I, Drakulich S, Dikici E, Tomic-Canic M, Deo S, et al. Nanotechnology-driven therapeutic interventions in wound healing: Potential uses and applications. ACS Central Science. 2017;**3**:163-175. DOI: 10.1021/acscentsci.6b00371

[45] Jack AA, Nordli HR, Powell LC, Powell KA, Kishnani H, Johnsen PO, et al. The interaction of wood nanocellulose dressings and the wound pathogen *P. aeruginosa*. Carbohydrate Polymers. 2017;**157**:1955-1962

[46] Poonguzhali R, Khaleel Basha S, Sugantha KV. Novel asymmetric chitosan/PVP/nanocellulose wound dressing: In vitro and in vivo evaluation. International Journal of Biological Macromolecules. 2018;**112**:1300-1309

[47] Rees A, Powell LC, Chinga-Carrasco G, Gethin DT, Syverud K, Hill KE, et al. 3D bioprinting of carboxymethylated-periodate oxidized nanocellulose constructs for wound dressing applications. BioMed Research International. 2015;**2015**:1-7

[48] Azuma K, Izumi R, Osaki T, Ifuku S, Morimoto M, Saimoto H, et al. Chitin, chitosan, and its derivatives for wound healing: Old and new materials. Journal of Functional Biomaterials. 2015;**6**:104-142. DOI: 10.3390/jfb6010104

[49] Taheri A, Mohammadi M. The use of cellulose nanocrystals for potential application in topical delivery of hydroquinone. Chemical Biology & Drug Design. 2015;**86**:102-106

[50] Meng LY, Wang B, Ma MG, Zhu JF. Cellulose-based nanocarriers as platforms for cancer therapy. Current Pharmaceutical Design. 2017;**23**:5292-5300

[51] Edwards J, Fontenot K, Liebner F, Condon B. Peptide-cellulose conjugates on cotton-based materials have protease sensor/sequestrant activity. Sensors. 2018;**18**:2334

[52] Edwards J, Fontenot K, Liebner F, Pircher N, French A, Condon B. Structure/function analysis of cotton-based peptide-cellulose conjugates: Spatiotemporal/kinetic assessment of protease aerogels compared to nanocrystalline and paper cellulose. International Journal of Molecular Sciences. 2018;**19**:840

[53] Dong S, Cho HJ, Lee YW, Roman M. Synthesis and cellular uptake of folic acid-conjugated cellulose nanocrystals for cancer targeting. Biomacromolecules. 2014;**15**:1560-1567

[54] Shi X, Zheng Y, Wang G, Lin Q, Fan J. pH- and electro-response characteristics of bacterial cellulose nanofiber sodium alginate hybrid hydrogels for dual controlled drug delivery. RSC Advances. 2014;**4**:47056-47065

[55] Cheginia SP, Meamar R, Varshosaz J. Co-delivery of venlafaxine and doxycycline by films of cellulose nanofibers for diabetic foot ulcers. Pharmacy Update. 2018;**1**:36-37

[56] Samadian H, Zamiri S, Ehterami A, Farzamfar S, Vaez A, Khastar H, et al. Electrospun cellulose acetate/gelatin nanofibrous wound dressing containing

berberine for diabetic foot ulcer healing: In vitro and in vivo studies. Scientific Reports. 2020;**10**:8312

[57] Singla R, Soni S, Patial V, Kulurkar PM, Kumari A, Padwad MS, et al. Cytocompatible anti-microbial dressings of *Syzygium cumini* cellulose nanocrystals decorated with silver nanoparticles accelerate acute and diabetic wound healing. Scientific Reports. 2017;7:10457

[58] Zheng Z, Liu Y, Huang W, Mo Y, Lan Y, Guo R, et al. Neurotensin-loaded PLGA/CNC composite nanofiber membranes accelerate diabetic wound healing. Artificial Cells, Nanomedicine, and Biotechnology. 2018;**46**:1-9

[59] Tong WY, bin Abdullah AYK, binti Rozman NAS, bin Wahid MIA, Hossain MS, Ring LC, et al. Antimicrobial wound dressing film utilizing cellulose nanocrystal as drug delivery system for curcumin. Cellulose. 2018;**25**:631-638

[60] Singla R, Soni S, Patial V, Markand Kulurkar P, Kumari A, Mahesh S, et al. In vivo diabetic wound healing potential of nanobiocomposites containing bamboocellulose nanocrystals impregnated with silver nanoparticles. International Journal of Biological Macromolecules. 2017;**105**:45-55

[61] Bajpai SK, Ahuja S, Chand N, Bajpai M. Nano cellulose dispersed chitosan film with AgNPs/curcumin: An in vivo study on Albino rats for wound dressing. International Journal of Biological Macromolecules. 2017;**104**:1012-1019

Chapter 6

# Application of Metal-Organic Framework as Reactive Filler in Bisphenol-A-Based High-Temperature Thermosets

*Vijayakumar Chinnaswamy Thangavel, Siva Kaylasa Sundari Saravanamuthu, Arunjunai Raj Mahendran and Shamim Rishwana Syed Mohammed*

## Abstract

Excellent thermoset monomers, bisphenol-A-based biscyanate ester (BADCy) and bispropargyl ether (BPEBPA), are synthesized and thermally cured to high-temperature thermosetting polymers. The nanoporous aluminum fumarate (Al_FA_A), an interesting Metal-Organic Framework (MOF), is synthesized in an eco-friendly manner and used as a reactive nanoparticle filler. The interaction of fumarate π bonds (*trans* -CH=CH-) in MOF with the reactive end functional groups (-O-C ≡ N) in cyanate ester (CE) and ($-CH_2-C \equiv CH$) in bispropargyl (BP) ethers is focused in these hybrid nanocomposites. The % decrease in enthalpy of curing in the organic and the inorganic blends (~60% for CE and ~ 10% for BP) indicates the interaction exciting between the MOF and the organic component. The addition of the aluminum fumarate MOF increases the glass transition temperature of the polymers. The amount of heat released for every increase in 1°C during the temperature window of curing ($\Delta H_c/T_E-T_S$) of the neat BADCy resin is approximately 2.4 times higher than the blend (BADCy+Al_FA_A). But BPEBPA shows only a 1% higher temperature curing window compared to its blend with MOF. The metal hotspots present in the hybrid nanocomposites may be the reason for the decrease in the thermal stability, and the % char residue is noted at 700°C. The TG-FTIR studies are done to predict the gaseous products ($CO_2$) evolved during thermal degradation.

**Keywords:** biscyanate ester, bispropargyl ether, aluminum fumarate MOF, thermal properties, TG-FTIR

## 1. Introduction

The cyanate ester (CE) and the acetylene terminated (AT) resins are specialty materials and find applications in several fields. Of which, the bisphenol-A-based dicyanate (BADCy) and the bispropargyl (BPEBPA) resins are advanced thermoset materials due to the formation of triazine [1] (polycyanurate) from BADCy [2] and products such as polyene, phenolic allenes by Claisen rearrangement, and chromene from aromatic bispropargyl ether (BPEBPA). The polycyanurate materials have high strength and toughness, high glass transition temperature, low water absorption, low dielectric constant, and good adhesion to a variety of substrates [3], and they combine the advantages of epoxies, fire resistance of phenolics, and high-temperature performance of polyimides. Several authors studied thermal degradation [4], curing kinetics, and blends of bismaleimides with bispropargyl ethers [5–8]. The authors already have experience in synthesizing the bispropargyl ether with several swivel groups. The most important study is copper-salt-assisted polymerization of bispropargyl of bisphenol-A [9] and the blend of propargyl ether with bismaleimide [10].

Polymer blending with other thermosets (or) with inorganic salts/materials [10, 11] has been studied by different research groups. This created a large scope for investing in the materials through resin modification and reinforcements. The authors work toward the influence of chain extension in BMIM composites and studied the effects of chain extensions on thermal and mechanical properties [12]. In their investigation, Siva Kaylasa Sundari *et al.* [13] studied the influence of nanoporous aluminum fumarate in cyanate ester matrix. It was reported that the double bond of aluminum MOF was involved in triazine formation. The synthesis, characterization, and thermal degradation kinetics by the model free approach of aluminum fumarate were already presented [14]. This study compares the bulk polymerization and polymer thermal properties having different functional group systems (bisphenol-A-based cyanate ester and bispropargyl ether) with aluminum fumarate MOF as particulate reinforcements. The structures of bisphenol-A-based monomers are presented in **Figure 1**.

The exponential increase in the design and development of Metal-Organic Frameworks (MOFs) is due to their tailorable properties and unprecedented functionality. The MOFs have a wide range of advantage and extend applications in several smart fields [15, 16] and also in biomedical applications [17, 18]. MOFs are formed by strong covalent bonds between inorganic metal salts and organic linkers. The author already studied the detailed aspects of synthesis and procedure optimization of aluminum

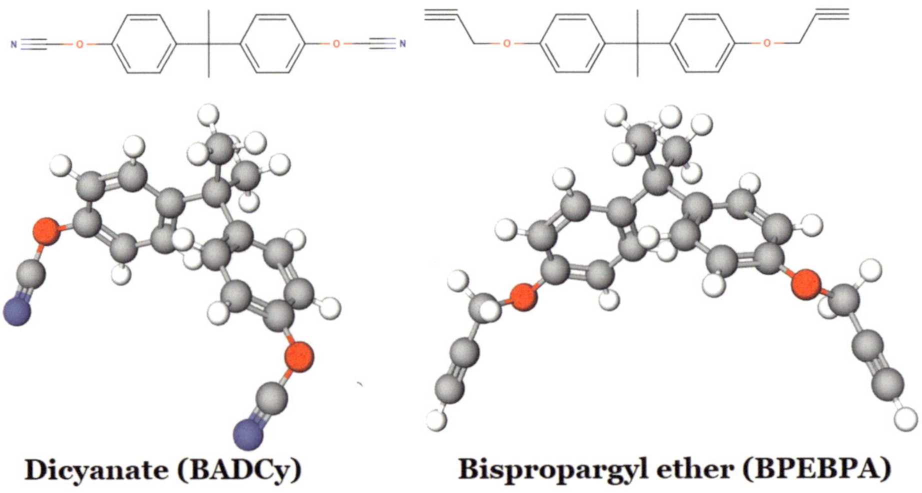


**Figure 1.**
*Structures of bisphenol-A-based monomers.*

fumarate and its isothermal model free [14] and model fitting kinetics. The formation of MOF-polymer composites can be obtained by the in-situ polymerization reaction, MOF construction using polymeric ligands, post-synthetic grafting of polymers onto the reactive ligands, the post-synthetic introduction of pre-formed polymers and MOFs self-assemble around polymers [19]. Markedly, this invention relates to the solvent-free process for preparing the MOF-polymer composite by the post-synthetic grafting of pre-formed polymers on reactive MOFs. The polymerized material contains allenes, phenolic allenes by claisen rearrangement, chromenes from aromatic propargyl ether, and also the linear polyene. The involvement of fumarate π-bonds in MOFs with reactive double bonds of polymerized bispropargyl provides a complex bulk and random hybrid polymer brushes, which makes the material applicable for functionalized applications.

## 2. Materials and methods

### 2.1 Materials

The compounds bisphenol A dicyanate (BADCy), bispropargyl ether (BPEBPA) and the nanoporous aluminum fumarate MOF (Al_FA_A) were synthesized. Bisphenol-A, sodium hydroxide, tetra butyl ammonium bromide, and isopropanol were obtained from Merck India Ltd., Mumbai, India, and were used without any further purification. Propargyl chloride obtained from Grauer & Weil (India) Limited, Vapi, India, was distilled (Boiling point: 57°C) before use. The synthesis procedure and the characterizations of bisphenol-A dicyanate [13] and the nanoporous aluminum fumarate MOF [14] were reported in our previous work.

#### *2.1.1 Synthesis of bispropargyl ether of bisphenol-A (BPEBPA)*

A mixture of 0.2 mol of bisphenol-A, 200 mL of 20% aqueous sodium hydroxide (NaOH), and 0.01mol of tetrabutyl ammonium bromide was kept at room

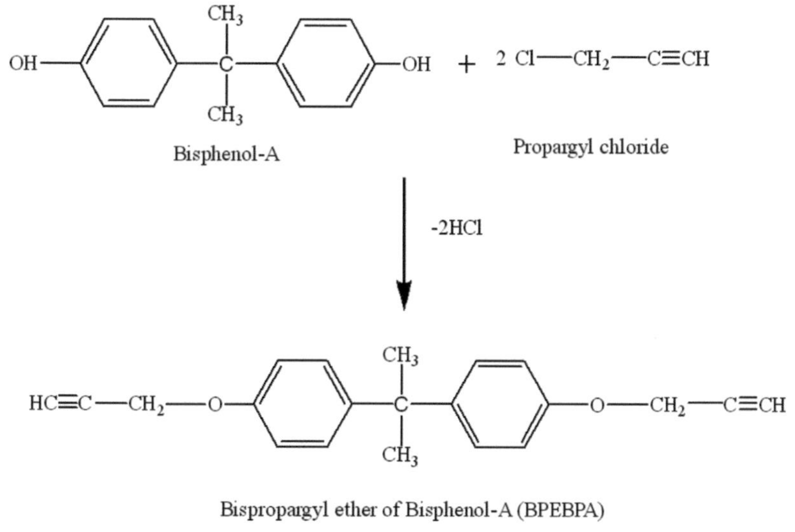


**Figure 2.**
*Synthesis of bispropargyl ether of bisphenol-A.*

temperature. About 0.6 mol of propargyl chloride was added over 10 min to the above mixture with constant stirring, and the resultant reaction mixture was stirred for 16 h. The resulting white crystals were filtered and washed twice with 200 mL of water and then with 50 mL of isopropanol. This results in a yield of 95% of BPEBPA. The reaction for the synthesis is presented in **Figure 2**.

### *2.1.2 Preparation of hybrid nanocomposites*

The dried BADCy/BPEBPA and the Al_FA_A (99:1, w/w) were taken in an agate mortar, and the mixture was ground repeatedly to have effective mixing. The mixture

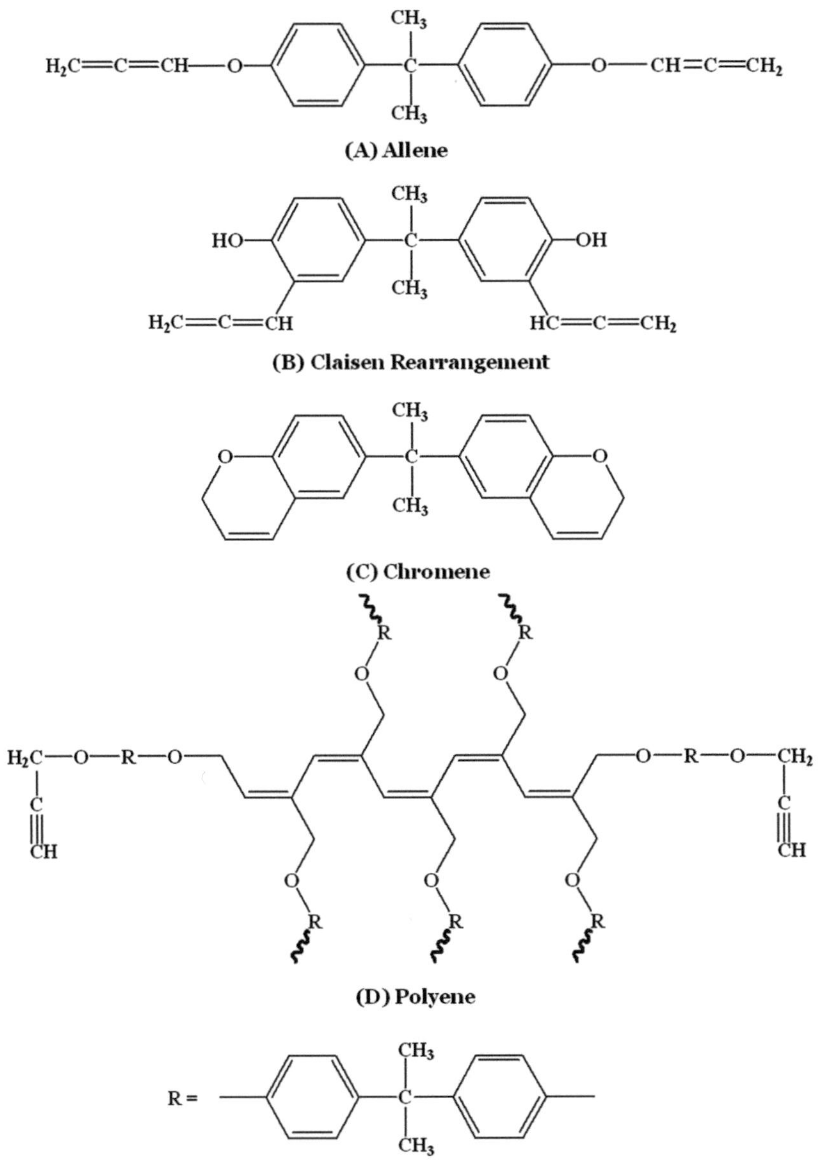


**Figure 3.**
*The products formed during the polymerization of bisphenol-A-based bispropargyl ether (BPEBPA).*

was then dried and preserved for polymerization. The pure BADCy/BPEBPA and their blends with nanoporous aluminum MOF were taken in separate borosilicate glass test tubes and flushed with dry oxygen-free-nitrogen gas and polymerized at 240°C for 6 h. After the polymerization, the samples were allowed to cool to room temperature, removed from the test tubes, ground to a coarse powder, packed, and stored for further analysis. The reaction mechanism for the involvement of fumarate in triazine formation is presented in our previous work [13]. The possible products during the polymerization of bispropargyl ether are presented in **Figure 3**.

### 2.2 Methods

The Fourier Transform Infrared Spectrum (FTIR) of the material was recorded on a Fourier transform infrared-8400S spectrophotometer, Shimadzu, Japan, using the potassium bromide disc technique. The differential scanning calorimetric (DSC) curves were recorded using TA Instruments DSC Q20 using a non-hermetic aluminum pan. The samples were heated from ambient to 350°C at 10°C $min^{-1}$ in a nitrogen atmosphere (50 mL $min^{-1}$). The thermal degradation behavior of the material was examined using TGA Model Q50 supplied by TA Instruments, Waters India Pvt., Ltd., Bengaluru – 560,086, India. The measurements were carried out using approximately 3–4 mg of the sample in a high-purity nitrogen atmosphere, and the flow of nitrogen to the balance area was 40 mL $min^{-1}$, and the sample was swept with a nitrogen flow of 60 mL $min^{-1}$. The obtained TG and DTG curves were analyzed using the universal analysis 2000 software provided by TA instruments. The TG-FTIR study of polymerized samples was carried out in a TA Instruments TGA Q5000 V3.10 Build 258 at a heating rate of 10°C $min^{-1}$ from ambient to 800°C. Nearly, 3–4 mg samples was used for the analysis in the nitrogen atmosphere (balance flow: 10 mL $min^{-1}$ and sample flow: 25 mL $min^{-1}$), and the FTIR spectra of the volatiles formed during the thermal degradation were recorded every 30 s.

## 3. Results and discussion

### 3.1 Aluminum fumarate

The Fourier Transform Infrared Spectroscopy was done for Al_FA_A to find the existence of intermolecular hydrogen bonding between the layers of aluminum fumarate. The Al_FA_A shows an onset degradation temperature at 410°C, attains a maximum at 484°C, and ends at 561°C. The material Al_FA_A is stable up to 400°C, and the amount of residue noted at 700°C is 33%. The crystallinity of Al_FA_A is 89% (Scherrer equation), and the average mean particle diameter of 50 nm. The Brunauer-Emmett-Teller (BET) surface area of Al_FA_A was found to be 937 $m^2g^{-1}$; monolayer volume ($V_m$) of 215.21 $cm^3g^{-1}$; total pore volume ($V_{pore}$) of 0.38 $cm^3g^{-1}$; and a mean pore diameter of 1.63 nm. The thermal degradation behavior of the aluminum fumarate by model free kinetics [14] and the curing behavior of BADCy and BADCy+Al_FA_A were already presented [13].

### 3.2 FTIR studies

The FTIR spectra for synthesized and polymerized materials are given in **Figure 4**, and the bands were reported in **Table 1**. The FTIR bands for the pre and

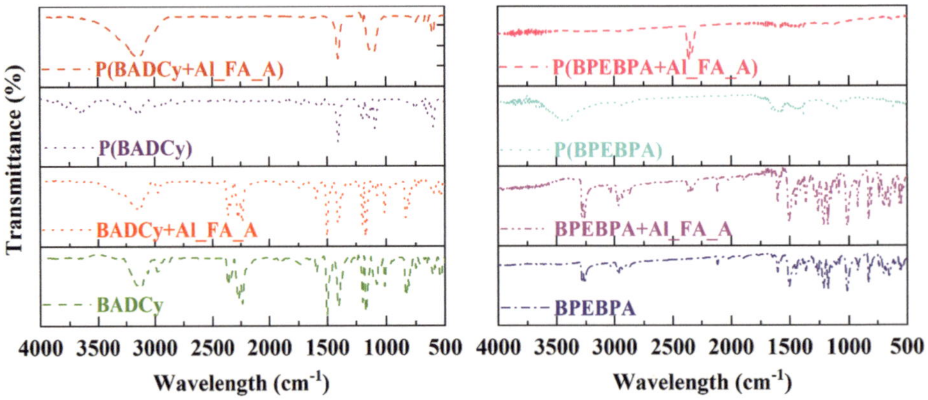


**Figure 4.**
*FTIR spectra for monomers and polymers.*

| S.No. | Frequency ($cm^{-1}$) | Bands (BADCy) | Bands (BPEBPA) |
|---|---|---|---|
| 1. | 3259 | — | ≡C-H stretching |
| 2. | 3100 | Phenolic -OH stretching | — |
| 3. | 2970 | C-H stretching in $CH_3$ | C-H stretching in $CH_3$ |
| 4. | 2120 | — | C ≡ C stretching |
| 5. | 2240 to 2260 | C ≡ N stretching | — |
| 6. | 1510 | C=N-C stretching of triazine ring | — |
| 7. | 1400 | O-C=N stretching of triazine ring | — |
| 8. | 1100 to 1200 | C-O-C stretching | C-O-C stretching |
| 9. | 820 | C-H out of plane deformation | C-H out of plane deformation |

**Table 1.**
*The frequency and corresponding bands of cyanate ester and bispropargyl system.*

post-polymerization of BADCy and BADCy+Al_FA_A at different temperatures were discussed in detail in our previous work [13]. The compound BPEBPA shows a characteristic band at 2120 $cm^{-1}$ for C ≡ C stretching, 3259 $cm^{-1}$ for ≡C-H stretching, and 700$cm^{-1}$ for ≡C-H deformation. The blending of fumarate MOFs in BPEBPA showed similar spectra to monomers. The polymerization that happened via the claisen polymerization was confirmed by the presence of the ketone group in IR spectra. The propargyl/bispropargyl involved in allene rearrangement then follows the claisen rearrangement to form keto-allene, which undergoes enolization for phenolic allene. The keto-allene further reacts to form the chromene structure [8], which shows a band at 1630 $cm^{-1}$ and a cyclic ether group at 1150 $cm^{-1}$.

### 3.3 DSC studies

The DSC thermograms for the monomers and the blends are given in **Figure 5**. The details regarding the melting and curing behavior values for materials are given in **Table 2**.

The BADCy shows the onset curing temperature ($T_S$) at 196°C, the temperature at which the curing rate is maximum ($T_{max}$) at 252°C, and the endset temperature

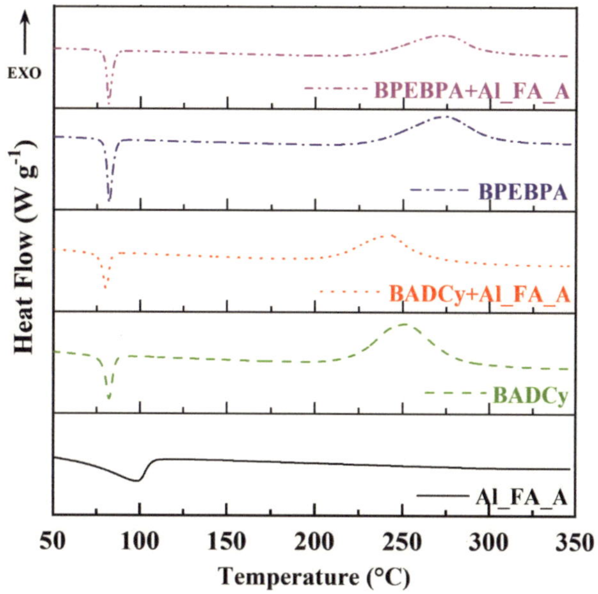


**Figure 5.**
*DSC thermograms of monomers and blends.*

| Sample Code | $T_m$ (°C) | $\Delta H_f$ (J g$^{-1}$) | $T_S$ (°C) | $T_{max}$ (°C) | $T_E$ (°C) | $T_E$-$T_S$ (°C) | $\Delta H_C$ (Jg$^{-1}$) | $\Delta H_C$/ $T_E$-$T_S$ (Jg$^{-1}$°C$^{-1}$) |
|---|---|---|---|---|---|---|---|---|
| BADCy | 82 | 31 | 196 | 252 | 334 | 138 | 211 | 1.53 |
| BADCy+Al_FA_A | 80 | 16 | 192 | 243 | 329 | 137 | 86 | 0.63 |
| BPEBPA | 82 | 45 | 205 | 274 | 332 | 127 | 207 | 1.63 |
| BPEBPA+Al_FA_A | 81 | 43 | 211 | 273 | 335 | 124 | 184 | 1.48 |

**Table 2.**
*Parameters from DSC curves.*

($T_E$) at 334°C. The addition of nanoporous material decreases the $T_S$ to 192°C, $T_{max}$ to 243°C and the $T_E$ to 329°C. The BPEBPA shows the $T_S$ at 205°C, the $T_{max}$ at 274°C, and the $T_E$ at 332°C. But the addition of Al_FA_A increases the $T_S$ to 211°C for BPEBPA. The enthalpy of curing (ΔHc) for BADCy is 211 J g$^{-1}$, whereas incorporation of Al_FA_A in BADCy drastically decreases the value to 86 J g$^{-1}$, which amounts to a decrease of 59%. The addition of aluminum fumarate in bisphenol-A-based thermoset decreases the enthalpy of fusion (50%) for BADCy and shows a negligible change in the BPEBPA matrix. The addition of aluminum fumarate increases the glass transition temperature ($T_g$) of polymer.

The enthalpy of combustion decreases by 60% for BADCy and by 10% for BPEBPA systems. The reduction in the $\Delta H_f$ and $\Delta H_c$ values for the blend indicates the interaction existing between the organic and inorganic components in the blend. The $T_E$-$T_S$ values of both the pure resins and the blends show an almost similar rate of curing. The amount of heat released for every increase in 1°C during the temperature window of curing ($\Delta H_c$/$T_E$-$T_S$) of the neat BADCy resin is approximately 2.4 times higher than the blend (BADCy+Al_FA_A). But BPEBPA shows only a 1% higher-temperature curing window compared to its blend with MOF. From the DSC curves, a profound effect was observed during the physical blending of aluminum fumarate in bisphenol-A-based thermosets.

### 3.4 TG studies

The thermal analyses of the polymerized samples are presented in **Figure 6**. The polymers are less stable than the nanoporous aluminum MOFs. The details regarding the onset and endset degradation temperatures, $T_E$-$T_S$ value, mass loss (%) and the residue (%) at 700°C are given in **Table 3**. Compared to the BADCy, the BPEBPA matrix shows broad degradation. The influence of nanoporous material does not affect the degradation pattern in BADCy, but it shows a drastic difference in the BPEBPA polymer. The effect of copper salts (CuCl: cuprous chloride; $CuCl_2.2H_2O$: cupric chloride dihydrate; $CuSO_4.5H_2O$: copper sulfate pentahydrate; $CuCO_3.Cu(OH)_2$: basic copper carbonate; $(CH_3COO)_2Cu.H_2O$: cupric acetate monohydrate; CuO: cupric oxide) in the BPEBPA matrix reported that the addition of copper salts decreases the onset degradation temperature [9]. Similarly, the addition of Al_FA_A decreases the thermal stability in both cyanate (BADCy) and the bispropargyl (BPEBPA) systems. The limiting oxygen

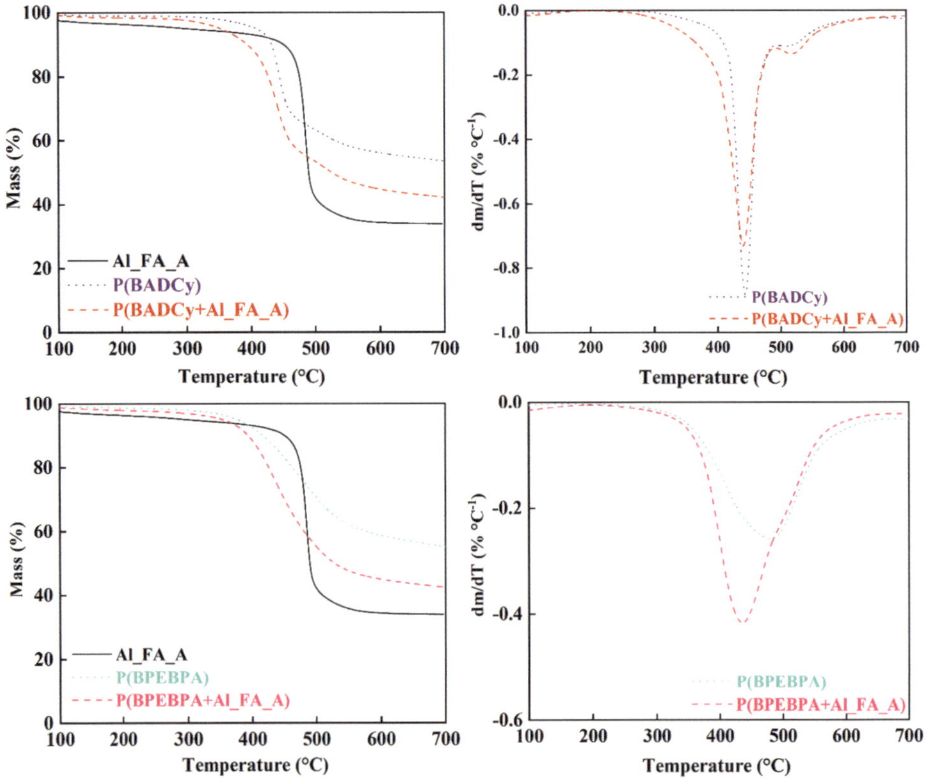


**Figure 6.**
*TG and DTG curves of polymers.*

| Sample Code | First Degradation Stage | | | | | Second Degradation Stage | | | | Residue (%) at 700°C | LOI |
|---|---|---|---|---|---|---|---|---|---|---|---|
| | $T_s$ | $T_{max}$ | $T_E$ | $T_E$-$T_S$ | Mass Loss (%) | $T_s$ | $T_{max}$ | $T_E$ | Mass Loss (%) | | |
| Al_FA_A | 410 | 484 | 548 | 138 | 65 | — | — | — | — | 33 | 0.31 |
| P(BADCy) | 376 | 443 | 623 | 247 | 45 | — | — | — | — | 53 | 0.39 |
| P(BADCy+Al_FA_A) | 327 | 439 | 658 | 331 | 55 | — | — | — | — | 42 | 0.34 |
| P(BPEBPA) | 220 | 476 | 672 | 452 | 42 | 672 | 751 | 797 | 4 | 52 | 0.38 |
| P(BPEBPA+Al_FA_A) | 206 | 435* | 662 | 456 | 55 | — | — | — | — | 40 | 0.34 |

**Overlapped degradation.*

**Table 3.**
*Degradation parameters of Al_FA_A and polymers.*

index (LOI) for the nanoporous MOF and the polymers was calculated using Van Krevelen method [20] eq. (1).

$$LOI = \frac{17.5 + (0.4 * Char\ Residue)}{100} \tag{1}$$

### 3.5 TG-FTIR studies

The 3D TG-FTIR spectrum of aluminum fumarate is presented in **Figure 7**. The major compounds released from the Al_FA_A were water ($H_2O$), carbon

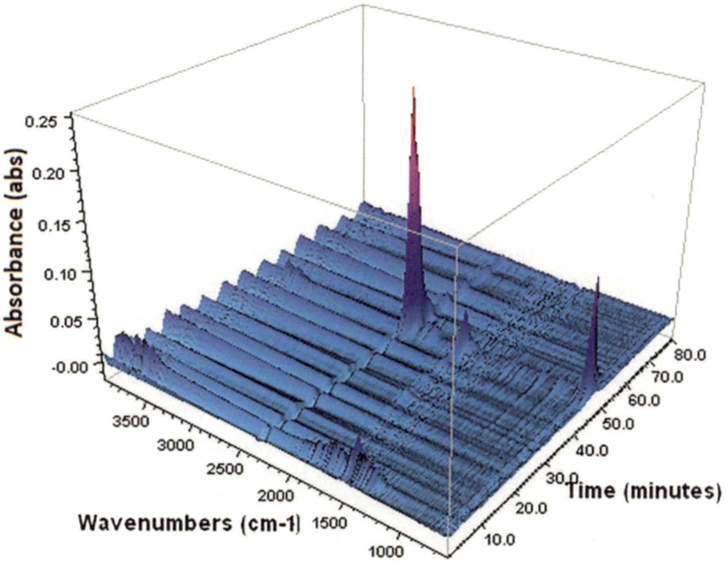


**Figure 7.**
*3D: TG-FTIR spectra of aluminum fumarate.*

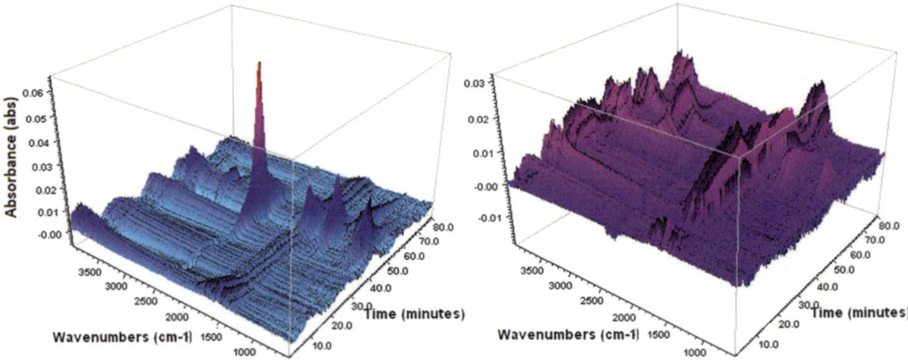


**P(BADCy+Al_FA_A)** **P(BPEBPA+Al_FA_A)**

**Figure 8.**
*3D: TG-FTIR spectra of polymer blends.*

dioxide ($CO_2$), carbon monoxide (CO), and acetylene. A detailed discussion regarding the bond dissociation energies of the aluminum fumarate is provided in the previous work [14]. The 3D TG-FTIR spectra for polymer blends are presented in **Figure 8**.

Compared to P(BADCy) and P(BADCy+Al_FA_A), the polymers P(BPEBPA) and P(BPEBPA+Al_FA_A) release more carbon dioxide ($CO_2$) at 2350 $cm^{-1}$ at all temperatures. The intensity of $CO_2$ evolution in P(BADCy) and P(BPEBPA) decreases due to the addition of nanoporous MOF, it indicates that the $CO_2$ can be captured in the tunnels of the MOF. Vinayagamoorthy *et al.* [6] and Dhanalakshmi *et al.* [9] studied the off-line pyrolysis and the evolved gas analysis technique of the bispropargyl ethers, and they concluded that the obtained gaseous degradation products should be able to liberate by cleavage at an activated benzylic (or) propargylic site in BPEBPA system. The thermal, mechanical, and electrical effects of Al MOFs in the di/trifunctional epoxies of epoxy resin composites were studied in our previous work [21, 22].

## 4. Conclusion

The nanoporous aluminum fumarate (Al_FA_A), bisphenol-A-based cyanate ester (BADCy), and bispropargyl ether (BPEBPA) were synthesized. The high-temperature thermoset polymers were obtained from the monomers and blended with Al_FA_A. The addition of aluminum fumarate MOF decreases the onset curing temperature ($T_S$) for BADCy and increases $T_S$ for the BPEBPA system without affecting the melting point. The influence of π-fumarate bonds of Al_FA_A in the polymer matrix was observed from the drastic changes in the enthalpy of fusion and enthalpy of combustion values. The onset thermal degradation temperature of polymers was decreased by the influence of MOF. The influence of nanoporous material does not affect the degradation pattern in BADCy, but it shows a substantial difference in the BPEBPA matrix. The major products released from the polymers are water ($H_2O$), carbon dioxide ($CO_2$), carbon monoxide (CO), and acetylene.

## Acknowledgements

The authors express their sincere thanks to the Management and Principal of Kamaraj College of Engineering and Technology (Autonomous), S.P.G.C. Nagar, K. Vellakulam-625701, Madurai, India, for providing all facilities to do the work.

## Funding

No funding was received for this study.

## Conflict of interest

The authors have no relevant financial or non-financial interests to disclose.

### Author statement

Supervision, Writing – Review and Editing: Dr. C. T. Vijayakumar.
Methodology, Writing – Original draft preparation: S. Siva Kaylasa Sundari.
Validation: Dr. M. Arunjunai Raj and.
Validation: Dr. S. Shamim Rishwana.

## Author details

Vijayakumar Chinnaswamy Thangavel[1*], Siva Kaylasa Sundari Saravanamuthu[2], Arunjunai Raj Mahendran[2] and Shamim Rishwana Syed Mohammed[3]

1 Thangavel Match Industries, Tamil Nadu, India

2 Wood K Plus, Competence Center for Wood Composites and Wood Chemistry, St. Veit an der Glan, Austria

3 Department of Chemistry, Kamaraj College of Engineering and Technology (Autonomous), Tamil Nadu, India

*Address all correspondence to: ctvijay22@yahoo.com

## References

[1] Hammerton I. Chemistry and Technology of Cyanate Ester Resin. Dordrecht: Springer; 1994

[2] Reams JT, Guenthner AJ, Lamison KR, Vij V, Lubin LM, Mabry JM. Effect of chemical structure and network formation on physical properties of di(cyanate ester) thermosets. ACS Applied Materials & Interfaces. 2012;**4**(2):527-535

[3] Sheng X, Akinc M, Kessler MR. Creep behavior of bisphenol E cyanate ester/alumina nanocomposites. Materials Science and Engineering: A. 2010;**527**(21-22):5892-5899

[4] Dhanalakshmi JP, Raj MA, Vijayakumar CT. Thermal degradation kinetics of structurally diverse poly(bispropargyl ethers-bismaleimide) blends. Chinese Journal of Polymer Science. 2016;**34**(3):253-267

[5] Liu F, Li W, Wei L, Zhao T. Bismaleimide modified bis propargyl ether bisphenol a resin: Synthesis, cure, and thermal properties. Journal of Applied Polymer Science. 2006;**102**(4):3610-3615

[6] Vinayagamoorthi S, Vijayakumar CT, Alam S, Nanjundan S. Structural aspects of high temperature thermosets – Bismaleimide/propargyl terminated resin system-polymerization and degradation studies. European Polymer Journal. 2009;**45**(4):1217-1231

[7] Liu F, Li W, Wei L, Zhao T. Blended resins based on a new propargyl-functional resin: Synthesis, cure, and thermal properties. Journal of Applied Polymer Science. 2006;**102**(5):4207-4212

[8] Dhanalakshmi JP, Pitchaimari G, Alam S, Vijayakumar CT. High-temperature matrix resins based on bispropargyl ethers (BPEs) – Part 2. High Performance Polymers. 2014;**26**(7):691-702

[9] Dhanalakshmi JP, Alam S, Vijayakumar CT. High temperature matrix resins based on bispropargyl ethers – Part 1. High Performance Polymers. 2012;**25**(4):416-426

[10] Dhanalakshmi J, Vijayakumar CT. Copper salt–assisted polymerization of bispropargyl ether–bismaleimide blend. Journal of Elastomers & Plastics. 2018;**51**(1):52-63

[11] Zhao Z, Lei B, Du W, Yang Z, Tao D, Tian Y, et al. The effects of different inorganic salts on the structure and properties of ionic liquid plasticized starch/poly(butylene succinate) blends. RSC Advances. 2020;**10**(7):3756-3764

[12] Sivaprakash S, Siva Kaylasa Sundari S, Shamim Rishwana S, Dhanalakshmi J, Vijayakumar CT. Low-temperature processable glass fiber reinforced aromatic diamine chain extended bismaleimide composites with improved mechanical properties. Polymer Composites. 2022;**43**(10):6987-6997

[13] Siva Kaylasa Sundari S, Shamim Rishwana S, Ramani R, Arunjunai Raj M, Vijayakumar CT. Particulate reinforcements in Dicyanate composites with Nanoporous aluminum fumarate as reactive filler: Thermal properties. Polymer (Korea). 2022;**46**(4):1-10

[14] Siva Kaylasa Sundari S, Shamim Rishwana S, Kotresh TM, Ramani R, Indu Shekar R, Vijayakumar CT. Effect of structural variation on the thermal degradation of nanoporous aluminum

fumarate metal organic framework (MOF). Journal of Thermal Analysis and Calorimetry. 2021;**147**(8):5067-5085

[15] Rowsell JL, Yaghi OM. Strategies for hydrogen storage in metal--organic frameworks. Angewandte Chemie (International Ed. in English). 2005;**44**(30):4670-4679

[16] Chowdhury MA. Metal-organic-frameworks for biomedical applications in drug delivery, and as MRI contrast agents. Journal of Biomedical Materials Research. Part A. 2017;**105**(4):1184-1194

[17] Hoskins BF, Robson R. Design and construction of a new class of scaffolding-like materials comprising infinite polymeric frameworks of 3D-linked molecular rods. A reappraisal of the zinc cyanide and cadmium cyanide structures and the synthesis and structure of the diamond-related frameworks [N(CH3)4][CuIZnII(CN)4] and CuI[4,4′,4″,4″'-tetracyanotetraphenylmethane]BF4.xC6H5NO2. Journal of the American Chemical Society. 2002;**112**(4):1546-1554

[18] Henschel A, Gedrich K, Kraehnert R, Kaskel S. Catalytic properties of MIL-101. Chemical Communications (Camb). 2008;**35**:4192-4194

[19] Yang S, Karve VV, Justin A, Kochetygov I, Espín J, Asgari M, et al. Enhancing MOF performance through the introduction of polymer guests. Coordination Chemistry Reviews. 2021;**427**. Article number: 213525

[20] Van Krevelen DW, Te Nijenhuis K. Properties of Polymers. 4th ed. Amsterdam: Elsevier; 2009. p. 1032

[21] Siva Kaylasa Sundari S, Shamim Rishwana S, Ramani R, Vijayakumar CT. Improvement in electrical and mechanical properties of di/trifunctional epoxies-based hybrid composites having metal organic frameworks (MOFs) as nanoparticulate filler. MRS. Communications. 2022;**12**:250-256

[22] Ganeshan J, Jeyadevi S, Siva Kaylasa Sundari S, Arunjunai Raj M, Pitchaimari G, Vijayakumar CT. Thermal, mechanical and electrical properties of difunctional and trifunctional epoxy blends with nanoporous materials. Journal of Elastomers and Plastics. 2022;**54**(3):494-508